Metropolitan College of NY
Library - 7th Floor
60 West Street
New York, NY 10006

The Construction of Human Kinds

BD
450
.M2646
2016

The Construction of Human Kinds

Ron Mallon

UNIVERSITY PRESS

Metropolitan College of NY
Library - 7th Floor
60 West Street
New York, NY 10006

OXFORD
UNIVERSITY PRESS

Great Clarendon Street, Oxford, OX2 6DP,
United Kingdom

Oxford University Press is a department of the University of Oxford.
It furthers the University's objective of excellence in research, scholarship,
and education by publishing worldwide. Oxford is a registered trade mark of
Oxford University Press in the UK and in certain other countries

© Ron Mallon 2016

The moral rights of the author have been asserted

First Edition published in 2016

Impression: 1

All rights reserved. No part of this publication may be reproduced, stored in
a retrieval system, or transmitted, in any form or by any means, without the
prior permission in writing of Oxford University Press, or as expressly permitted
by law, by licence or under terms agreed with the appropriate reprographics
rights organization. Enquiries concerning reproduction outside the scope of the
above should be sent to the Rights Department, Oxford University Press, at the
address above

You must not circulate this work in any other form
and you must impose this same condition on any acquirer

Published in the United States of America by Oxford University Press
198 Madison Avenue, New York, NY 10016, United States of America

British Library Cataloguing in Publication Data

Data available

Library of Congress Control Number: 2015960481

ISBN 978-0-19-875567-8

Printed in Great Britain by
Clays Ltd, St Ives plc

Links to third party websites are provided by Oxford in good faith and
for information only. Oxford disclaims any responsibility for the materials
contained in any third party website referenced in this work.

For Johnelle

Contents

List of Figures — ix

Introduction — 1

Part I. Constructing Human Kinds

1. Constructing and Constraining Representations: Was *Race* Thinking Invented in the Modern West? — 15
2. Constructing Categories: Concepts, Actions, and Social Roles — 48
3. Social Roles that Matter — 68
4. Representation and Moral Hazard — 94
5. Performance, Self-Explanation, and Agency — 111

Part II. Realizing Social Construction

6. Social Construction and Reality — 137
7. Achieving Stability — 162
8. Achieving Reference — 182

Part III. Conclusion

9. Alternatives and Implications — 207

Acknowledgments — 217
Bibliography — 219
Index of Names — 241
Index of Subjects — 245

List of Figures

0.1	Representation-world co-determination	1
3.1	Representation-world co-determination	82
3.2	Environmental construction	83
5.1	Intentional performance of "multiple personality"	123
6.1	World-to-representation epistemic constraint	140
6.2	Social forces as an alternative explanation	145
7.1	Social change produces Taylor Instability	165
7.2	Hacking instability and labeling effects	170

Introduction

We humans classify ourselves and each other into a dizzying array of *categories* including: *elderly* and *adolescent*; *male* and *female*; *gay* and *straight* and *bisexual*; *white* and *Asian* and *Latino* and *black*; *depressed* and *angry* and *happy* and *maudlin*; *schizophrenic* and *psychopathic* and *autistic* and *agoraphobic*; *diabetic* and *epileptic*; *stupid* and *smart*; *careless* and *careful*; *liberal* and *conservative*; *Buddhist, Catholic, Hindu, Protestant, Mormon, Jewish,* and *Muslim*. We attempt to represent these human categories in the course of our engagement in a whole range of enterprises including explaining and predicting the behaviors of other individuals and groups; signaling to and coordinating with others; representing the world to ourselves; and stigmatizing, valorizing, and regulating the behavior of ourselves and others.

My focus here is upon a certain kind of explanation of human categories, one that explains the existence or features of the category by appeal to our practices of representing it: a *social constructionist* explanation. Throughout the social sciences and humanities, this family of influential and provocative claims suggests that not just *representations* of human categories, but human *categories* themselves are socially constructed—are in some way a product of a community's practices of labeling and differentially treating category members. As a first pass, we can depict the constructionist position as a simple looping diagram (see Figure 0.1).

We expect that, when things go right, representation is constrained and explained by the way the world is (the bottom arrow in Figure 0.1), but constructionists appeal

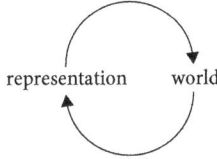

Figure 0.1 Representation-world co-determination

to the features of our representations to explain features or categories in those parts of the world the representation is about (the top arrow). Among the most interesting of these putative objects of construction are those that are apparently natural categories of person like sex categories (e.g. *male* or *female*) or race categories (e.g. *black, white, Asian*). By "apparently natural," I mean that there is a widespread tendency to treat the categories as *natural kinds*, membership in which is explained by possession of some natural (often understood as biological) property that also explains the typical features of category members. In contrast to such explanations, social constructionists offer a competing account of category membership or of the explanation for category-typical features in the mental states and social practices of a community that distinguishes the category.

Consider some influential examples of constructionist explanations. The critical sociologist Stuart Hall claims that:

'black' is essentially a politically and culturally *constructed* category, which cannot be grounded in a set of fixed trans-cultural or transcendental racial categories and which therefore has no guarantees in nature. (1996, 443)

And the legal scholar and feminist activist Catherine MacKinnon has maintained that:

Gender has no basis in anything other than the social reality its hegemony constructs. Gender is what gender means. The process that gives sexuality its male supremacist meaning is the same process through which gender inequality becomes socially real.
(1987, 173)

Similarly, the philosopher Michel Foucault (1978) has written that both representations of homosexuality, but also the category of homosexuality, emerged in nineteenth-century Europe:

We must not forget that the psychological, psychiatric, medical category of homosexuality was constituted from the moment it was characterized – Westphal's famous article of 1870 on "contrary sexual sensations" can stand as its date of birth – less by a type of sexual relations than by a certain quality of sexual sensibility, a certain way of inverting the masculine and feminine in oneself. Homosexuality appeared as one of the forms of sexuality when it was transposed from the practice of sodomy onto a kind of interior androgeny, a hermaphrodism of the soul. The sodomite had been a temporary aberration; the homosexual was now a species. (1978, 43)

Such social constructionist claims about human categories play a number of theoretical roles.

Because both social constructionists' and biological naturalists' accounts of a category C concede the existence of C, they contrast with *skeptical* accounts about C who claim that C does not exist. Such skepticism often motivates *eliminativism*, the view that recommends that we stop making use of "C" talk, that we stop purporting to refer to C. For instance, because social constructionists about race offer social (rather than biological) accounts of racial categories, they join with biological racial naturalists in asserting the reality of race in the face of racial skepticism. As in the race case, I interpret constructionists as asserting social explanations of categories that they take to really exist and that form defeasible grounds for resisting eliminativism.

Some constructionists are thus concerned with articulating the social reality that social practices can produce, but a constructionist focus on human categories also grows from ethical concern with the effects of our representational practices on questions of authority, freedom, and justice. Although we deploy many human representations effortlessly and without much thought, examination of our practices of classification reveals them to be exquisitely complex in their function, application, evaluative content, and causal effects. One reason we classify ourselves and others is suggested by my talk of explanation, prediction, and causal significance: we take our representations to pick out real, relevant, causally important differences among the persons we classify. And it is in part because these differences can be relevant to our theoretical and practical projects that we use these representations and information they carry as we make plans, form our identities, coordinate our actions, and modify our attitudes and policies towards ourselves and others.

But human beings are also famously political animals. Humans are centrally concerned with others. This leads us to be concerned with presenting ourselves to others as rational and moral beings (Haidt 2001). And it leads us to reason about ourselves and others as a means of getting things we want (Mercier and Sperber 2011). One way that we do both of these things—both self-presentation and social negotiation—is by creating and using representations of human kinds to justify, rationalize, and exculpate human actions, policies, attitudes, and traits of persons to ourselves and to others. Because representations often carry information about the way the world is, we can use them to persuade, manipulate, and regulate the behavior of representation-using creatures like us. This fact leads quickly to the concern that these representations (and the practices they in turn cause and rationalize) may be in some way mistaken, and also to the concern that such mistaken beliefs and practices may produce or perpetuate inequality and oppression—a concern that is itself a central feature of liberal and critical social philosophy. Another reason for a focus on constructionist explanation is, then, the hope that where our representational practices form part of the mechanism creating or sustaining differentiation of some human category, altering these practices may be a way of altering the existence or character of the category as well.

Here my focus is upon offering an account of how the social construction of real, causally significant human categories is possible, and upon drawing and exploring implications of such an account. My aim is not, except where attempted, to closely interpret the views of any specific constructionist author, nor is it an attempt to defend a social constructionist account of race or of any other particular human category. Rather, my overall aim is synthetic: I engage a range of disciplines in the course of articulating a central strand of social constructionist thought, that how we represent ourselves and other humans profoundly shapes our thought, behavior, and society. I consider and address issues and problems that constructionist ideas raise in a way that I take to be broadly sympathetic to some core constructionist aims, but also to fit with ideas, concerns, and objections that emerge from contemporary psychology and the philosophy of science. Like a number of other recent authors, I aim to develop social constructionist themes within a broadly naturalistic and realist framework.[1] The framework is naturalist and realist in both the methodological and the metaphysical senses that it takes science to be an enormously successful epistemic enterprise that teaches us about how the world is.[2] But it is social constructionist in the sense that it explains by appeal to those features of the natural world that include human minds, culture, social practices, and their products. I develop and defend the view that social constructions themselves are, in the relevant senses, real features of the world that our ordinary and social scientific discourse can and do make reference to.

1 What is Social Construction?

Social constructionist thought grows out of a range of different intellectual sources:

- Normative engagement by philosophers and social theorists with socially salient and causally powerful categories like race and sex.
- Discussions by evolutionary psychologists using "social constructionist" as a label for social scientists who explain human behavior, social practices, and institutions more or less exclusively by appeal to human decisions, culture, and the material environment.

---

[1] E.g. Andreasen 1998; De Block and Du Laing 2007; Tooby and Cosmides 1992; Griffiths 1997; Haslanger 2012; Kelly 2011; Machery and Faucher 2005a, b; Mallon and Stich 2000; Murphy 2006; Sperber 1996; Stein 1990, 1999; Wilson 2005.

[2] See Mallon 2013b; see Prinz 2007 for more discussion of naturalism.

- The sociological study of scientific knowledge and the response to it by philosophers of science.

Given the broad range of constructionist claims and sources, is there anything they have in common? Most commentators on social construction have emphasized the importance to constructionists of insisting upon the contingency of objects of construction.[3] Ian Hacking puts the point schematically, writing that many constructionists hold that if X is constructed then:

> X need not have existed, or need not be at all as it is. X, or X as it is at present, is not determined by the nature of things; it is not inevitable. (Hacking 1999, 6)

But how does saying that a thing need not have existed or is not inevitable distinguish objects of construction?

In some sense, it is possible that both natural objects like stars and artifacts like garters need not have existed, or might have been different than they are, had *some* facts been different. Similarly for alphabets, aquifers, cotton candy, cumulonimbus clouds, flint rock, glam rock, the PTA, RNA, video tapes, and the visual system: they might all have had different properties or not have existed given, for example, different physical laws or a different physical history of the universe. A denial of inevitability only makes sense against a background view about what it is that might have made the difference.

Social constructionists are particularly interested in phenomena that are contingent specifically upon human mental states or decisions or culture or social practices—contingent upon the theories, texts, conventions, routines, and conceptual schemes of particular individuals and groups of people in particular places and times. As a rough characterization, we can say that:

> *Social construction*: X is socially constructed if and only if X's existence or persistence or character is caused or constituted by human mental states, decisions, culture, or social practices.

I employ this very general characterization of social construction throughout this book—precisifying it where necessary—but letting its generality do work in capturing the wide range of discussions I engage.[4] It allows us to recognize the affinity of explicitly "constructionist" accounts with a wide range of work in the social sciences and humanities that abjures the label "social construction"—for example, talk by Foucault, Judith Butler, and others of "discursive formations,"

---

[3] E.g. Haslanger 1995/2012; Griffiths 1997; Hacking 1999; Kukla 2000.
[4] For useful discussions of sorts of social constructionism, see Haslanger 1995/2012 or Hacking 1999.

Hacking's discussions of "dynamic nominalism" and "historical ontology," Arnold Davidson's work on "historical epistemology," as well as a host of titles that discuss "inventing," "creating," or "making up" various apparently natural phenomena.[5] It also includes a wide range of more traditional and naturalistic work in, for example, history or anthropology, that adverts to social, historical, or cultural explanations but without the use of "construction" language.

2 Representations and Categories

Above, we noted two sorts of social constructionist explanatory targets: *representations* and *categories*. By "representations," I mean the attitudes, theories, narratives, concepts, models, pictures, norms, rules, utterances, inscriptions, and texts by which we represent the world and structure our actions. By "categories," I mean the things those representations are about: the properties or kinds in the world that we refer to and theorize about.[6] By "kinds," I pick out those categories that are or could be relevant or important or significant enough to figure in our successful theories. Given the interest relativity that pervades our theoretical activities, including our assessments of what is relevant or important or significant or successful, some category's status as a "kind" in this sense is bound to be relative to this or that theory or explanatory interest too. I mostly leave these complications to the side. But I do note that some properties are objective enough (*has an odd number of hairs*) but are not important to us, and others (*has the varicella zoster virus*) structure successful theoretical and practical activities. Many will fall in between. By this criterion, many targets of constructionist analysis—race, gender, and homosexuality among them—are plausibly regarded as genuine kinds.

3 Constructing Representations

Representational construction involves the determination of a *representational attitude*, where in an individual or group comes to be related to a certain representational vehicle with a certain meaning. Philosophers often discuss *propositional attitudes*—attitudes like "believing" or "desiring" toward some proposition. For example,

---

[5] Butler 1990; Davidson 2001; Foucault 1972; Hacking 1986; Laqueur 1990.

[6] "Category" is ambiguous in philosophy: "sometimes it refers to *worldly kinds*; sometimes it refers to *mental words or concepts*" (Devitt 1991, 243). E.g. Griffiths (1997, 175f) uses it in the former way (as I do here); Khalidi (2013) in the latter.

Believing that the porch will need to be painted again next year.
Desiring that one not catch the SARS virus.

In attitude reports, we relate a person to a proposition. We might say, for instance:

Zahra believes that the train will be here soon.

We thereby express the proposition to which Zahra is related by uttering the representation "the train will be here soon."

But here I talk about constructing "representations," instead of simply sentences or propositions, in order to draw attention to the many more types of meanings and vehicles than propositions and sentences that may be the objects of constructive processes. And I talk of "attitudes" instead of simply "beliefs" in order to leave open the possibility that other attitudes may be constructed. Some theorists are concerned with the "construction of desire" (e.g. McKenna 1984). But social construction might also result in certain non-belief but belief-like entities. For instance, constructive processes may fix implicit attitudes (like Tamar Gendler's "aliefs"), or they might fix group beliefs.[7]

Representations take many forms in the mind and in human social life, but for the most part, I leave these complexities aside in what follows. In discussing the construction of representations, I primarily consider cases in which beliefs, belief-like attitudes, or the linguistic expressions of these attitudes are causally explained by human culture or decisions or social practices. Such work aims, in part, to explain why some representations (rather than others) come to be the vehicles of individual and collective acceptance (rather than other attitudes).

4 Constructing Categories

Categories are properties or kinds that may be instantiated in the world, and here we are especially interested in properties or kinds instantiated by humans. Category constructionists characteristically hold a number of positions regarding the relationship of the social construction of representations to the social construction of categories:

- *Dual Constructionism*: Social constructionists about human categories typically are also social constructionists about the representations of those categories. So, for instance, constructionists about gender categories typically also think that our ideas, or concepts, or theories of gender (and their representational vehicles) are social constructions.

---

[7] Gendler 2008; see Pettit 2003 on groups with mental states.

- *Tandem Construction*: Social constructionists about human categories often hold that the construction of a category occurs *in tandem* with the construction of the representation of a category. For instance, according to Foucault, the category of homosexuality appeared along with the articulation of the representation.
- *Representational Control*: Belief in tandem construction seems to be a consequence of a third feature of constructionist accounts of categories, one that holds that categories are constructed *via* our representations of them. The representation figures as a crucial part of the mechanism by which the category is constructed. It follows from this conception that if we could intervene upon those representations, we could alter the construction of the category.
- *Necessary-Description Category Construction*: Many constructionists seem to hold that category invention occurs in tandem with the emergence of representations *with a particular content or meaning*. So for instance, in the passage above, Foucault holds that the category of homosexuality did not appear until a representation of a category characterized by an underlying "quality of sexual sensibility" came to be articulated. This suggests that the availability of a new way of thinking about or describing a human category may have itself been crucial to the construction of the category itself.
- *The Construction of Human Kinds*: Many category constructionists are concerned with categories like race, gender, or illness categories that are themselves salient and causally powerful features of the world; these categories plausibly count as kinds—features of the world whose significance makes them candidates for inclusion in our best theories.
- *Revelatory Aim*: Many category constructionists have a *revelatory aim*: they aim to show that a putatively natural category or trait is actually caused or constituted by unappreciated human decision or culture or social practice.
- *Successful Reference*: category constructionists hold that human representations refer to the categories that they play a role in constructing. For Foucault, "homosexual" refers to a category that came into existence in the nineteenth century. For MacKinnon, gender terms pick out a social reality that they themselves structure.

A central aim of this book is to offer a "how possibly" account of the social construction of human categories on which many of these positions can be consistently co-realized, given what we know about the social and natural world. I begin my theorizing with the thought that putatively natural but socially constructed categories would need to be real and causally powerful, and I proceed to try to understand how this could be so.

5 Overview

This book is divided into two main parts and a conclusion. The chapters in Part I are concerned with articulating an account of the social construction of human kinds based on social roles. Part II develops the idea that a naturalistic, realistic approach to social construction is possible and plausible. Part III relates this account to other approaches to, and concerns with, human categories.

5.1 Part I

Constructionist claims about representations and about categories have an asymmetric plausibility. On the one hand, that our human category representations are at least partially the result of a process of cultural construction is enormously plausible. We know that the stock of human category representations changes over time with some representations emerging (homosexual, programmer, astronaut) and others receding nearly into oblivion (serf, pharaoh, cooper, oracle, manciple, reeve). Human category representations also vary across cultures and among individuals. On the other hand, the idea that putatively natural categories (understood as kinds of things in the world) might be products of our thought and action is surprising, and it demands further explanation. How could sex or gender or race or emotion or sexual orientation be a product of our social practices while simultaneously being widely thought to be natural?

In Chapter 1, I begin by *qualifying* the enormously plausible claim; I argue that human category representations are not exclusively the product of social-constructive forces and the state of the world. Thus, in contrast to the ubiquitous Dual Constructionism among category constructionists, I argue for the possibility that some core features of human category representations should be given partially nonconstructionist explanations, resulting in a hybrid constructionist approach (Mallon and Kelly 2012). I then go on, in Chapters 2–5, to offer an articulation and in-principle defense of the more implausible claim that categories are constructed.

Chapter 1 asks whether racial representations are a novel invention in modern times. Drawing upon evolutionary-cognitive explanations of essentialist representations, I suggest that essentialist construals of human groups emerge across a wide range of cultures because of a cognitive predisposition to think about some groups in essentialist ways. More generally, this example shows that the social construction of representations of human categories may be qualified by appeal to psychological constraints without lending credence to the content of the category representations.

In Chapters 2 and 3, I turn to offer an account of the social construction of human categories more generally that focuses upon some account of how our practices of representing can produce categories that are causally significant. In Chapter 2, I understand the social construction of human categories as the production of *social roles*, by the creation of common knowledge human category representations in a culture. In Chapter 3, I build upon this framework with an account of some psychological and environmental mechanisms by which social roles may produce, constitute, or sustain causally significant kinds—what I call *entrenched social roles*.

In Chapter 4, I turn to explore evidence for a psychological connection between representations of categories as natural and moral exculpation. I suggest such a connection is empirically and theoretically plausible, and I argue that it illuminates some constructionist concern with the content of our representations of human categories. Then, in Chapter 5, I discuss *performative* constructionist accounts that appeal to the influence of widespread category representations upon intentional behavior by the agents occupying the categories (e.g. Hacking 1995a; Butler 1990). Such explanations, together with the constructionist commitment to revelation, raise a *puzzle of intention and ignorance*: Why doesn't the intentional performance of a category undermine its putative naturalness? A plausible answer is to appeal to a failure of self-explanation, and I suggest it is a failure to *locate* one's mental states in a causal explanation of one's thoughts and actions. I go on to argue that false self-explanations of this sort undermine our capacities for agency in nonobvious ways, and this further illuminates and justifies political concern with our representations of our natures.

5.2 Part II

In Chapters 6, 7, and 8, I relate constructionism to broader metaphysical questions.

I begin in Chapter 6 by observing that the possibility of a naturalistic and realistic treatment of social construction is sometimes obscured by the volatile intellectual and political disputes surrounding the so-called "science wars." Since work by Thomas Kuhn, Paul Feyerabend, and others, a growing body of research has rightly emphasized the need for empirically adequate—and therefore *social*—accounts of actual scientific practice.[8] This attention to practice, in turn, has led to further discussions apparently calling into question the rationality of scientific

---

[8] Kuhn 1970/1962; Kuhn et al. 2000; Feyerabend 1975.

practice and the reality of the world the sciences study. The result has been that the term "social construction" is widely associated with some sort of radical anti-realism.

In Chapter 6, I set out some of the central, anti-realist strands of social constructionist argument and situate my alternative, metaphysically moderate and realist approach to the construction of human representations and categories among them. In Chapter 7, I respond to suggestions by social theorists and constructionists to the effect that the culturally produced structure of social reality is so unstable as to undermine the possibility for knowledge of the sort that is enjoyed by the natural sciences. I suggest that in some circumstances social categories can achieve stability, and may even come to be *more* tightly coupled to our successful theories than natural kinds. In Chapter 8, I take up Successful Reference, arguing that human kind terms can successfully refer to constructed categories in the face of widespread misunderstanding about the character of those categories.

5.3 Part III

I conclude in Chapter 9 by contrasting the entrenched social role account of category construction with several alternate approaches to human categories, and I draw some conclusions about the normative import of covert constructionist work, including its import for social change.

As the saying goes, "hope is a good breakfast, but a bad supper." It's time to turn to the details.

PART I

Constructing Human Kinds

1

Constructing and Constraining Representations

Was *Race* Thinking Invented in the Modern West?

The capacity to represent particular categories of human in certain ways—categories like cosmonaut or diabetic or reeve—arises in particular cultural contexts with the articulation of representations that express certain contents. What about the capacity to represent race? Most people (at least in contemporary European-American cultures) think racially.[1] That is, they cognitively and verbally represent humans as divisible into racial kinds, and they draw further inferences and associate further properties of a variety of sorts with membership in these kinds.

Why do we do this? Why do we think racially? Why do our racial representations have the specific content that they do? Why do they seem to license the specific inferences that they do?

Much work in the humanities (including in philosophy) and the social sciences seems to assume that racial representations are to be explained exclusively via some sort of social constructionist explanation. On this approach, our racial representations are to be understood by considering their cultural predecessors, the historical and institutional context of their emergence, and the theoretical and practical choices people have made regarding how to represent humans as members of racial groups. In fact, the emerging consensus has been that the contemporary concept or idea of race was invented sometime in the modern era, perhaps as recently as the nineteenth century (see, e.g., Banton 1977; Banton and Harwood 1975; Fredrickson 2002; Guillaumin 1995;

---

[1] In this chapter, "America" designates the United States of America.

Hannaford 1996; Smedley and Smedley 2012). The historian George Fredrickson writes:

> It is the dominant view among scholars who have studied conceptions of difference in the ancient world that no concept truly equivalent to that of 'race' can be detected in the thought of the Greeks, Romans, and early Christians. (2002, 17)

To be sure, Fredrickson and others recognize that many similar forms of representation predate genuinely racial thinking, but they nonetheless hold that a significant change in the content or meaning of people's thinking about human groups emerged in recent centuries. I will call this the *Conceptual Break Hypothesis*.

> *Conceptual Break Hypothesis*: Sometime in or since the Renaissance, some fundamental change occurred in the European and American tradition of thinking about the human groups that we now call "races"—a substantial change in the concept, meaning, or theory by which people represent those groups.

On some renderings the Conceptual Break Hypothesis involves specifically a claim of a failure of conceptual identity among older and newer concepts of human groups.[2] On such a view, contemporary racial terms came to have a fundamentally different meaning than preceding terminology, such that it is a mistake to take previous authors to mean the things that we do with terms like "race."

Just what is this fundamental change in concept, meaning, or theory? This question is important because we would like to know whether the Conceptual Break Hypothesis is true. But it is also important because understanding constructionist claims about the invention of concepts will be important to my later discussion of the constructionist explanations of human categories. Since social constructionists typically view the construction of human categories as crucially involving ways of representing the category (what I called "Representational Control" in the Introduction), one way of assuring the social construction of a human category is via an account on which the social construction of a representation is necessary to constitute the category (via what I called "Necessary-Description Category Construction" in the Introduction).

I argue that a plausible candidate element for both interpreting what social constructionist defenders of the Conceptual Break Hypothesis say, and for vindicating the truth of the Conceptual Break Hypothesis, is the idea that in recent centuries individuals in the European-American cultural tradition began to conceive of race in an *essentialist* manner. That is, they came to believe that members of the groups that they now think of as races differed not only in

---

[2] Puzzo 1964, 579; Hannaford 1996, 6; Fredrickson 2002, 17; Smedley and Smedley 2012, 13ff.

superficial properties like skin color, hair type, or body morphology, but also in possession of some perhaps unseen property that

(1) was (at least causally) necessary and sufficient for membership in the race;
(2) explained typical features of the race; and
(3) was passed on from parents to children.

Call this view *racial essentialism*. At least some defenders of the Conceptual Break Hypothesis claim that racial essentialism is a culturally and historically local product of modern European and American thought.

But is this true? Did people begin to engage in racial essentialism in thinking about human populations or cultural groups only in modern times, perhaps as a result of the scientific revolution?

I argue that the answer is "no." While social constructionist accounts of racial representations have it that essentialist thinking about race is a relatively recent phenomenon, some evolutionary cognitive work tells a different story, one on which essentialist thinking about human groups is itself, or is a product of, a psychological mechanism that is *innate, domain-specific,* and *species-typical* (Gelman 2003; Hirschfeld 1996; Sperber 1996; Gil-White 2001a, 2001b; Machery and Faucher 2005a, 2005b; Jones 2009).[3] In saying a mechanism is "innate," I mean to indicate that these evolutionary cognitive theorists hold at least that the trait is largely *culturally canalized* or *invariant*—that it develops relatively invariantly across a wide range of cultures.[4] To say that a mechanism is "domain-specific" is to say that, unlike *domain general* cognitive capacities (like memory, attention, or perception) that are employed across a wide range of problem domains, this mechanism is specialized for solving a narrower problem or problems. This is not, however, to say that the mechanism is for thinking about race. Evolutionary cognitive theorists that consider the adaptive function of such cognition do not typically believe that there is a *race module* or a specific mental mechanism designed for thinking racially. Rather, they understand racial essentialism to be the by-product of a mental mechanism that is for

---

[3] While I characterize this research program as "evolutionary cognitive," some theorists give only passing attention to evolutionary concerns (e.g. Gelman 2003) though others (e.g. Tooby and Cosmides 1992; Hirschfeld 1996; Keil 1992; Gil-White 2001a) more fully engage the project of using evolution to consider what (if any) the adaptive function of a trait might be.

[4] This is usually considered a necessary, but not sufficient condition for innateness (see, e.g., Mallon and Weinberg 2006). Such developmental stability suggests that the developmental resources for producing the trait are available in a wide range of cultures, either because they are part of the genotype and other internal resources of the developing organism, or because they are generally available in environments in which humans can survive. Recent critiques of the concept of innateness (e.g. Griffiths 2002) do not deny the existence of developmentally invariant traits in this sense, only that they are perspicuously labeled "innate."

something—or some things—else.[5] And to say that they are "species-typical" is to say that, like having two arms and legs, or eyes, or ears, or hair, these cognitive capacities are traits that humans usually possess.[6]

To this characterization, I add two caveats. First, to say that a trait is innate, canalized, or invariant is not to say that it is unchangeable. These terms describe a regularity in the outcome of the process by which a trait develops or is acquired. In the present context, what matters is the existence of a strong degree of developmental invariance across a range of cultural environments.[7] But, as we will see, such invariance is consistent with the process or processes that produce a trait being susceptible to interruptions that alter their outcomes. Second, to say that some aspects of contemporary racial representations are developmentally invariant across cultures, or are the result of a mechanism that is relatively developmentally invariant across cultures, is not to say that all of them are. Rather, "racial essentialism" of this sort can manifest in a range of different ways in different cultures (for example as medieval European anti-Semitism or as Indian *caste*), and these ways may be more or less important given the social and cultural milieu. We revisit these two caveats below as we note some ways that racial essentialism varies in its manifestation or even fails to manifest. For now, I emphasize that this explanation of racial essentialism does not imply that essentialist or other racial beliefs are unchangeable, let alone that racism is.

Returning to our overarching question, what seems clear is that the sort of account of racial essentialism emerging from the evolutionary cognitive psychological literature is at odds with understanding the content of a conceptual break as essentialist. Evolutionary cognitive accounts suggest that essentialist thinking should emerge relatively robustly across human cultures and history while many social constructionists hold it to be a culturally local product of the modern era. It seems that if one is true, then the other is false. Because the historical emergence of racial essentialism is invoked in defense of a conceptual break, this evolutionary cognitive program gives us some reason to doubt the truth of the Conceptual Break Hypothesis as well, a consequence that would force a reconsideration of a considerable literature discussing

---

[5] This view is at least partially a consequence of their racial skepticism: evolutionary cognitive theorists typically do not believe that biological races (or racial essences) exist. It would therefore be odd for them to suppose a cognitive mechanism was adapted to tracking them.

[6] Some, especially in the philosophy of biology, worry that the anti-essentialism central to Darwinian thinking about biology (e.g. Mayr 1976; Sober 1980) undermines evolutionary psychological claims of species-typical traits as well. This worry seems to me ill founded, in part for reasons given by Machery (2008).

[7] Thus for present purposes, it would be okay if (as recent work on Bayesian learning algorithms intimates) *domain-general learning* mechanisms account for the acquisition of folk essentialist cognitive structure, so long as the acquisition occurs relatively invariantly across cultural contexts.

culture and racial classification, and ultimately, a reconsideration of classifications of other human categories as well. This reason is only prima facie since it remains a possibility for defenders of the Conceptual Break Hypothesis to spell out the content of the break in some other way (something that some already do).

Here's how my argument proceeds. In section 1, I show that one plausible way to interpret and vindicate the Conceptual Break Hypothesis is to hold that the belief that human peoples, nations, or populations possess a distinctive, explanatory essence emerged in the modern era. Then, in section 2, I review work by developmental psychologists that supports the surprising claim that racial essentialism emerges from or is substantially constrained by innate, domain-specific, and species-typical cognitive mechanisms. In Section 3, I turn to consider some cross-cultural evidence for this same conclusion. In Section 4, I consider objections to the evolutionary cognitive account of racial essentialism. And in section 5, I conclude by considering options for defenders of the Conceptual Break Hypothesis, and the possibility of some other, nonconceptual break.

1 In What Does the Conceptual Break Consist?

1.1 What is a change in concept, meaning, or theory?

The Conceptual Break Hypothesis posits some change in the concept, meaning, or theory by which people represent the human groups that we now call "races." What does such a change consist in? Because of their attention to changing beliefs or theories over time, it is plausible to think that social constructionist authors assume that substantial changes in widespread belief or theory amount to changes in the meanings of key terms within the theory—an assumption that also motivates a *descriptivist* family of philosophical accounts of meaning that hold the meaning of a term is given (at least in part) by the role it plays in the belief set, conception, or theory in which it figures (e.g. Jackson 1998).[8] Roughly, on such a view, two terms will have the same meaning when they play the same theoretical role or figure in the expression of the same beliefs.

Some familiar lessons follow from these descriptivist philosophical accounts: if we allow *any* difference in theory or belief to be meaning constitutive, then conceptual breaks will be ubiquitous. Even two people within a single culture are likely to differ in *some* of the beliefs that they express with the same term. If I believe that the man in front of the coffee shop this morning was of a different

---

[8] Mallon and Stich (2000) interpret constructionist work as assuming some sort of descriptivist account. Cf. Chapter 2, Section 1.

race than me, and you are indifferent as to whether that is true, the meaning of "race" will differ for the two of us. Such *holistic* theories of meaning have the consequence that none of us are ever using our terms with the same meaning.

A plausible amendment to this approach is to pursue some *molecular* theory of meaning—to distinguish some subset of beliefs, call them the *core* beliefs, that constitute the meaning of the term. (Within this alternative approach, we can further distinguish relatively *opulent* accounts from more *austere* ones by *how many* beliefs they include in the core.) On this molecular approach, two persons or cultures will share the same belief just in case (or perhaps, to the extent that) those *core* beliefs are identical. The limitation to the core makes possible the sharing of the relevant beliefs. But the challenge for a molecular approach is to determine which beliefs are core and which are *peripheral* to meaning, a notoriously difficult task, and one that, since W. V. O. Quine's attack on the analytic-synthetic distinction, many philosophers regard as impossible.[9]

Fortunately, this is not a task we need to undertake here. Instead, we simply note that on a molecular construal of the relevant terms, the meaning of one term will be nonequivalent to that of another term whenever

(a) there is some difference in between the sets of beliefs in which the terms figure, and
(b) that difference occurs in the "core" or meaning constitutive beliefs for either term.

One might deny the Conceptual Break Hypothesis by insisting that Fredrickson and other defenders are alighting upon insignificant or nonmeaning-constitutive differences (thus denying b). If so then one has the difficult task of nontendentiously demarcating the core meaning of "race." But my own strategy is to allow, at least provisionally, that Fredrickson and many other scholars do identify a significant or core element of racial thinking. I instead deny the Conceptual Break Hypothesis by disputing (a). I deny that the candidate element they have identified is really absent in earlier and alternative conceptual schemes. Whether the element is really meaning-constitutive is thus irrelevant to the present argument.

1.2 Historical emergence of racial essentialism

What is this element? Some important defenders of the Conceptual Break Hypothesis suggest that it is the idea of racial essentialism. These defenders endorse:

---

[9] Quine 1953. Fodor and Lepore (1992) offer an illuminating analysis of these and related issues.

The historical emergence of racial essentialism (HERE) hypothesis: racial essentialism is a culturally specific and historically recent way of thinking about some human groups.

The HERE hypothesis is compelling for two general reasons. First, it makes historical sense given the emergence of the scientific revolution in post-Renaissance Europe and America, and of the emergence of the scientific study of humanity in the eighteenth and nineteenth centuries. Consider, for example, repeated attempts by prominent naturalists to divide humans into distinctive natural groupings and to explicate the nature of those groupings (e.g. Buffon 1791, Gobineau 1999).

Second, other social constructionist studies of human categories like homosexuality suggest that the emergence of essentialist thinking (e.g. Davidson 1990, 2001; Foucault 1978) was a quite general product of this scientific turn. Recall (from the Introduction) Michel Foucault's influential claim that the homosexual (both the idea and the kind itself) emerged in the nineteenth century as homosexual behavior came to be thought of as the product of a different underlying category of person. The HERE hypothesis fits nicely with other claims of the historical emergence of essentialist thinking for other categories.

The HERE hypothesis is also compelling as an interpretation of some defenders of the Conceptual Break Hypothesis. To begin with, some theorists endorsing the Conceptual Break Hypothesis explicitly suggest the HERE hypothesis. In addition, the existence of a racial essence is considered by many philosophers to be criterial for the modern *race* concept. It follows from this philosophical analysis that only after the idea of essence emerged could the contemporary concept *race* have existed. This conceptual claim provides a partially independent line of support for interpreting defenders of the Conceptual Break Hypothesis in terms of the emergence of essentialist thinking. That is: if the HERE hypothesis is true, and the conceptual claim is correct, the truth of the Conceptual Break Hypothesis would follow. In the remainder of this section, I argue for these interpretive claims in more detail.

1.3 A conceptual break in history

Consider first how some defenders of the Conceptual Break Hypothesis articulate the content of the shift. I begin with Michael Banton and Jonathan Harwood's 1975 book, *The Race Concept*. Banton and Harwood endorse the Conceptual Break Hypothesis, and go on to offer an account of what the break consists in. They write:

For thousands of years, writers have described men as members of particular families, clans, tribes: as citizens of cities and states, as belonging to peoples distinguished by their language, dress and customs. It was only in the nineteenth century that they started to describe men as belonging to races and to maintain that differences between men and peoples stemmed from race. It was then said that some men belonged to races which had an inborn tendency to expand their territory and rule over others who belonged to lesser strains and from whom less could be expected. (11)[10]

According to Banton and Harwood, the nineteenth century brought a new tendency to describe differences among individuals and groups in terms of inborn racial traits. More recent work by Banton continues in this vein, suggesting that in the nineteenth century, "the word 'race' came to signify a permanent category of humans of a kind equivalent to the species category" (1998, 6). According to this later passage from Banton, the emergence of the modern race concept occurs precisely as race comes to be conceived as species are—precisely as the kinds come to be thought of in a parallel way.

Similarly, French sociologist Colette Guillaumin endorses the Conceptual Break Hypothesis, claiming that the idea of race emerged at the beginning of the nineteenth century. She articulates the break as involving the emergence of "the *belief that groups are naturally diverse, because of endogenous characteristics which are determining factors in themselves, independently of history or economics*" (1995, 70-1; italics hers). Again, the idea is that some distinctive property of individual group members gives rise to the distinctive features of the groups they comprise.

Fredrickson's own account of the hypothesized conceptual break follows his discussion of a range of different sorts of ethnic discrimination in medieval Europe. He then proposes the following explanation of the conceptual break: "What was missing—and why I think such ethnic discrimination should not be labeled racist—was an ideology or worldview that would persuasively justify such practices" (24). Fredrickson goes on to elaborate the content of that worldview: "unless—or until—it was presumed that" a negative characteristic of a group "was organic and carried in the blood, it would not be proper to describe such an attitude as racist" (25). As with Banton, Harwood, and Guillaumin, Fredrickson holds that modern race arose with a conception of racial difference on which racial difference is caused by inborn properties of members of groups.

---

[10] There are a few apparently inconsistent strands in Banton and Harwood's presentation. In places, they present the race concept as dating from the late eighteenth or early nineteenth century (e.g. p. 7), and in other places they present it simply as having undergone an important shift at that date (e.g. p. 8, p. 13). The crucial thing for our purposes is that they endorse the Conceptual Break Hypothesis (whether or not the break is located within the concept of race), and that they explain it by appeal to a version of the HERE hypothesis.

In each of these cases, we find that the hypothesized change of content brought by the extension of the scientific revolution to the study of humans involved the idea that endogenous features of persons, typical of their race, explain racially typical phenotypic features, including not only features of their bodies (skin color, hair type, body morphology), but also the type and quality of the characters of members of a race, thus providing an individual level explanation for the differential "achievements" of races.[11]

1.4 Racial essentialism as criterial of race

The idea of racial essentialism has also played a pivotal role in recent philosophical discussions of race. While the term "essentialism" has a range of uses in the humanities, here the term seems to imply just the sort of idea that Banton, Harwood, Guillaumin, and Fredrickson are referring to: the idea that members of a race share a hidden trait, passed on in inheritance, that explains racially typical properties. In one recent characterization of the idea, K. Anthony Appiah writes of the *racialist* view that:

> we could divide human beings into a small number of groups, called "races," in such a way that the members of these groups shared certain fundamental, heritable, physical, moral, intellectual, and cultural characteristics with one another that they did not share with members of any other race. (1996, 54)

As it happens, contemporary philosophical debate is split over whether the sort of racial essentialism that Appiah writes of is constitutive of the race concept. Few versed in modern biology think that such essentialism is plausible for biological kinds (see, e.g., Mayr 1976; Sober 1980; Gannett 2001; Kitcher 1999; though see Devitt 2008). The recognition that an essentialist view of race is deeply mistaken has led philosophers of race toward one of two accounts of the role of essentialism in ordinary racial thinking.

Theorists that accept that the contemporary European-American concept of race necessarily involves such a commitment to essentialism tend to endorse racial skepticism (the view that race as ordinarily understood does not exist) (e.g. Zack 1993, 2002; Appiah 1995; Blum 2002; Glasgow 2009). But others, sympathetic to some variety of racial constructionism or biological realism (e.g. Andreasen 1998; Hardimon 2003; Haslanger 2003, 2005; Kitcher 1999; Mills 1998; Outlaw 1996; Root 2000; Sesardic 2003, 2010; Spencer 2014; Taylor 2000,

---

[11] Still other theorists hold that such essentialist ideas are part of the concept of race, whether or not they take these to constitute a conceptual break hypothesis (e.g. Smedley and Smedley 2012, 25–6; Barzun 1937).

2013) or pragmatism (Gannett 2010), deny that ordinary racial thinking necessarily involves racial essentialism.[12] They endorse a nonessentialist understanding of the content of ordinary race thinking. Crucially, these latter theorists need not deny that ordinary race thinking involves racial essentialism in the sense of Appiah's racialist (though they may); rather, they need simply deny that essentialism is constitutive of the contemporary meaning of "race" (see Mallon 2006).

This philosophical debate is independent of the Conceptual Break Hypothesis, and it looks to turn on a question of conceptual identity of the sort that I aim to avoid in the present discussion. However, the fact that racial essentialism has played a criterial role for the application of the race concept in many contemporary philosophical analyses provides a quasi-independent line of support for interpreting the Conceptual Break Hypothesis as vindicated by an emerging, essentialist view of races. For if a shift towards racial essentialist thinking did occur in the late eighteenth or early nineteenth century, and if constructionists are inclined to view such a change as significant to—or even criterial for—the application of the concept *race* in the way many philosophers have, then it would follow that they would come to endorse the Conceptual Break Hypothesis. So the recent philosophical focus on racial essentialism provides a plausible hypothesis for what the conceptual break consists in that dovetails nicely with what defenders of the Conceptual Break Hypothesis say.

To sum up where we are: the Conceptual Break Hypothesis asserts the historical particularity of racial thinking, or at least of modern, scientific, or biological racial thinking. In this section, I suggested that one plausible interpretation of the content of such a conceptual break is the HERE hypothesis. Such an explanation seems to flow out of what some defenders of the Conceptual Break Hypothesis say, but recent philosophical discussions of race suggest that it also has considerable plausibility as a claim about a significant (and perhaps constitutive) feature of racial thinking.

2 An Evolutionary Cognitive Account of Racial Essentialism

Much of the work we have been considering so far shares a social constructionist style of explanation, one that emphasizes the content of inherited culture and human decision in the explanation of the content of beliefs. Now I turn to

---

[12] As I understand it, racial constructionism is a kind of racial realism. See Part II.

consider a very different research tradition that has also focused upon the explanation of racial essentialism: an evolutionary cognitive research tradition. As I noted above, evolutionary cognitive accounts of representations explain the content of those representations by appeal to innate, species-typical, and domain-specific mechanisms (e.g. beliefs or biases in the acquisition, formulation, retention, or transmission of beliefs). In this section I briefly rehearse some of the developmental evidence for the early emergence of assumptions about the biological domain and about race. I begin by discussing some evidence for what I call *broad essentialism*, and then go on to sketch a more specific form of essentialism—*lineage essentialism*—that seems to be exhibited in the biological and racial domains.

2.1 Broad essentialism

For several decades, an ongoing evolutionary cognitive research program has investigated domain-specific, folk biological cognition (e.g. Carey 1995; Keil 1989, 1992; Springer and Keil 1989, 1991; Atran 1990, 1994, 1998; Atran and Medin 2008; Gelman and Markman 1986; Gelman and Wellman 1991; Gelman 2003).[13]

A related research program has investigated essentialism in cognizing human groups including races and ethnicities (e.g. Astuti et al. 2004; Hirschfeld 1996; Gil-White 2001a, 2001b; Machery and Faucher 2005a, 2005b; Kanovsky 2007; Jones 2009). One common (though not uncontested) way of expressing the findings (and the object of study) of this research is to say that it reveals that children are intuitive "essentialists," at least about biology, perhaps also about race, and possibly about a great many more domains. Summarizing her own interpretation of this extensive research program, Susan Gelman characterizes essentialism in a quite general way:

> what I mean by an "essence" is an underlying reality or true nature, shared by members of a category, that one cannot observe directly but that gives an object its identity and is responsible for other similarities that category members share. (2003, 4)

Gelman is characterizing what she takes the research to show about prescientific cognition of categories, namely that it operates with the assumption that a wide range of different kinds have "underlying" natural properties, such that

---

[13] The term "folk" here is used to pick out ordinary, prescientific cognition. While many evolutionary cognitive theorists suggest that parts of such folk theories will be species-typical, this is not entailed by the use of the term "folk." (There could, for example, be many folk biologies.)

(1) Possession of the property by an individual constitutes the individual's membership in the kind

and,

(2) Possession of the property by the members of the kind causally explains the typical properties of that kind.[14]

Call the combination of these two theses *broad essentialism*. Gelman and others have produced a range of evidence for the view that ordinary biological and racial cognition exhibit a commitment to such broad essentialism. Consider first that for both biological and racial kinds, people make judgments as if there is some property that, as Gelman says, "gives an object its identity" as a member of the kind. For instance, people judge that an individual that possesses the property remains a member of the kind even if its superficial properties are changed. In one study, Frank Keil showed that three–five-year-old preschoolers judged that while a costume can make a zebra appear to be a horse, the horse-costume-wearing zebra nonetheless remains a zebra (1989, chapter 11). In another study, Keil showed that by eight to ten years old, children judge that even scientists surgically altering a raccoon to make it seem like a skunk fail to change its kind identity (1989, chapter 9).[15]

Similarly, a distinction between having the typical properties of a kind and really being a member of the kind is also a much-discussed feature of American racial categories. That appearance is an indicator of race, rather than constitutive of it, is often discussed in connection with "passing" (cf. Piper 1992; Gooding-Williams 1988; Mallon 2004). Racial "passing" occurs when a member of one race seems to be a member of another, but it emerges because of the implicit assumption that kind-typical properties are merely indicative of, rather than sufficient for, racial membership. Such phenomena are complex and have diverse understandings in everyday life. Nonetheless, the idea of race as a discrete underlying property that may or may not be correctly indicated by appearance persists in ordinary racial thinking (e.g. Condit et al. 2004).

Gelman also mentioned that essences are taken to be explanatory, that they are "responsible for other similarities that category members share." In one early

---

[14] Gelman and Hirschfeld (1999) label the first "sortal essentialism" and the latter "causal essentialism." Motivation for "sortal essentialism" about natural kind terms also grows nicely out of a famous tradition in the philosophy of language and the philosophy of science that emphasizes possession of unseen microstructural properties as necessary and sufficient conditions for membership in a natural kind like "water" or "tiger" (Kripke 1980; Putnam 1975; Boyd 1991).

[15] Cf. Rothbart and Taylor (1992) on "inalterability."

study, Gelman and Ellen Markman (1986) offer subjects triads of animals consisting of a pair with a similar appearance (e.g. a blackbird and a bat) and an (overlapping) pair whose members were identified as belonging to the same kind category (e.g. a blackbird and a flamingo were each identified as birds). The goal was to see which sort of property (i.e. perceptual similarity or verbally labeled kind membership) subjects would use when projecting a novel, nonobvious property (e.g. "heart has a right aortic arch only"). Would subjects anticipate that the nonobvious property possessed by a bird was shared by the creature with a similar appearance (e.g. the bat) or the one identified as belonging to the same kind category (e.g. the flamingo)? Undergraduate subjects used the kind category on 86 per cent of the trials, and four year olds on 68 per cent. One way of explaining this sort of result is that subjects assume that bird kind membership is underwritten by an unseen property (a bird essence) that explains typical properties of birds (e.g. feathers, wings), and can be recruited as a basis for making a judgment about the novel property. (Cf. Gelman and Coley 1990; see Gelman 2003, chapter 2 for a review of this literature.)

Although the data may be more difficult to interpret, the same seems to be true in the case of race. There is some evidence that ordinary Americans understand race as grounded in underlying biology and perhaps genetics, and explain race-typical physical features in terms of biological or genetic differences (Condit et al. 2004; Sheldon et al. 2007). In comparison, explanation of other race-stereotypical features (e.g. differences in crime rates or academic ability) is now less regularly explained by appeal to biology and more often explained by environmental, cultural, or personal factors (Condit et al. 2004; Jayaratne et al. 2006; Jayaratne et al. 2009). This may reflect a broad, ongoing shift from more biological to more cultural explanations of racial difference (Schneider 2004, 466; Shulman and Glasgow 2010; Martin and Parker 1995), but also may be the product of complex social construals: for example, increasing sensitivity to seeming racist among American whites or perceived interconnections between biological explanations and moral evaluations (Jayaratne et al. 2009).[16] More generally, at times both scientific and colloquial racial theories have characterized racial types and used them to infer that individual members of a race instantiate the kind-typical properties. For example, when racial categories have been used to make inferences about group-typical sports ability, intellectual aptitude,

---

[16] While the overall implications of a shift to anti-essentialism are unclear (see, e.g., Shulman and Glasgow 2010; Mallon 2007a; Williams and Eberhardt 2008; Haslam et al. 2000), a number of studies have shown that belief in the fixedness of underlying traits as well as essentialist beliefs about categories are correlated with prejudice towards category members (Levy et al. 1998; Haslam et al. 2000; Haslam et al. 2006; Sheldon et al. 2007).

character, or criminality, this looks like an exercise of the same tendency underlying the use of biological categories to project properties to new instances.[17]

2.2 Inheritance thinking

Prescientific biological and racial cognition look to involve assumptions about underlying essences in this *broad essentialist* sense, but they also look to involve a distinct element—*inheritance thinking*—that looks to be shaped by some additional assumptions about kind-reproduction or lineage. Above we saw a commitment to the importance of lineage or inheritance in the evidence we reviewed to establish the plausibility of the HERE hypothesis. For example, Fredrickson (2002) talked of the belief that negative characteristics of a group are "carried in the blood" and Banton and Harwood (1975) talked about the emergence of ideas of races with "inborn tendencies" that could explain the differential achievements of peoples. I think a reading of these very different disciplinary traditions makes it plausible to think that in both cases they are assuming that racial kind-hood is passed along in biological reproduction. This tendency has been noted in the social philosophical literature on race as well, as when K. Anthony Appiah writes:

> There is from the very beginning until the present, at the heart of the system [of racial classification] a simple rule that very few would dispute even today: where both parents are of a single race, the child is of the same race as the parents. (1996, 77)

Inheritance thinking also seems to emerge in prescientific biological thinking. Here, studies of preschool- and kindergarten-age children seem to show early recognition that membership in biological categories is inherited from members of the same category. For example, young children judge that the children of members of a kind are also members of that kind (e.g. Johnson and Solomon 1997), just as in the case of racial categories.[18] And other work shows that being told that an individual baby animal or plant seed is of a kind causes children to judge that it will develop the typical properties of that kind, even in an "adoption" task that gives it a developmental environment characteristic of another kind. For example, four-year-old children judge that a baby cow, raised by pigs, will nonetheless say "moo" rather than "oink" (Gelman and Wellman 1991, experiments 3, 4, and 5). Extending this work to race, Lawrence Hirschfeld determined

---

[17] Which is not to say that is the only psychological disposition operating. For instance, Sarah-Jane Leslie emphasizes the importance of the projection of kind *atypical* properties to members of kinds, when those properties are of a certain sort (forthcoming).

[18] However, in the case of American racial categories, one can be black even if one has only one parent who is black (i.e. some form of a rule of "hypodescent" is observed in many American racial systems).

that children as young as four years of age hold human phenotypic features associated with race (skin color, hair color) to robustly develop in the face of cross-racial adoption (Hirschfeld 1995; though see Solomon 2002).

But what, exactly, are children assuming in these cases? Inheritance thinking seems to involve at least an implicit commitment to three causal claims:

The causal sufficiency of parent kind: Individuals in the offspring relationship to k's are k's.[19]

The causal necessity of parent kind: Genuine k's come only from other genuine k's.

The causal explanatoriness of k-hood: k's typically have certain properties because they are k's.

We see the first in, for example, a refusal to allow that certain sorts of transformations (e.g. costumes, or cross-species adoption) can confer a new kind membership on an individual *even when transforming its kind-typical properties* (though it remains possible that other, better, interventions might show that this is not a presupposition). And we see the second again in the insistence that superficially transformed individuals retain their original kind membership. We also see the second and the third in interspecies "adoption," and interracial adoption studies, in which individuals are held to have the categorization and kind-typical properties of their birth parents.[20]

Many theorists in the psychological literature pair broad essentialism and inheritance thinking in a straightforward way. They suggest that underlying these claims is the fact that kind members possess an essence that constitutes their membership in the kind. They further hold that what is passed on from parents to offspring is an essence which constitutes the offspring as members of the kind (see, e.g., Keil 1989; Gelman 2003; Ahn et al. 2001). This essence cannot (at least normally) be obtained in any other way. And this essence explains the kind-typical properties of kind members. Call this pairing of broad essentialism with inheritance thinking *lineage essentialism*. *Racial essentialism* (as we characterized it at the outset) is simply lineage essentialism about race.

The idea that ordinary, prescientific biological and racialist thinking might be structured by lineage essentialism is unsurprising. Talk of properties "carried in

---

[19] I here leave the "offspring" relation undefined so as to accommodate a range of views. As Kanovsky (2007) points out, the empirical evidence is consistent with a range of ways of acquiring essence.

[20] It could be that these first two commitments are the result of a folk commitment to deeper, modal claims of logical or metaphysical necessity—e.g. that a commitment to the causal necessity of parent kind is the result of a commitment to the logical necessity of parent kind.

the blood" and "inborn tendencies" lend themselves easily to lineage essentialist interpretation, as do subjects' willingness to project novel properties to new instances of a kind.

But lineage essentialism has special significance in the context of the discussion of evolutionary cognitive accounts of essentialism. In the background of some evolutionary cognitive research into essentialism in biological cognition is the idea that during human evolution cognitive adaptations that allowed humans to attend to and track stable features of the biological domain would plausibly have been selected for (e.g. Atran 1998; Keil 1992; Barrett 2005). In this context, assumptions about the inheritance of kind-hood or essences might be evidence of adaptation for biological thinking because some important biological entities (individuals and populations) reproduce themselves in ways such that projecting kind membership and kind-typical properties via inheritances is a useful epistemic practice (even if at the heart of the practice is an ontological assumption contemporary scientific biology largely rejects—the assumption of biological essence).[21] Similarly, since human populations and cultures also reproduce themselves, applying these assumptions to human groups may also be adaptive (Gil-White 2001a; Machery and Faucher 2005a).

2.3 Developmental evidence of theoretical sophistication

To say that adults and even children have ideas about race is one thing, but to say that they are culturally canalized is another. Evidence of stable, early development is one sort of evidence that is tapped to establish this stronger conclusion, as is cross-cultural evidence. But another source of evidence pits nativist hypotheses against competing, empiricist models of acquisition. Here I briefly consider two different sources of evidence that suggest cognition about human groups outruns knowledge gained from associations with salient visual differences. While this does little to undermine a representational constructionist model that emphasizes a rich tradition of cultural transmission, it does undermine a simpler empiricist model on which race thinking emerges smoothly as a generalization from visually salient differences.[22]

One source of evidence comes in Hirschfeld's (1996) extended argument that children's conceptual understanding of racial types seems to outstrip their ability to categorize people into races. For example, a range of data suggests that

[21] Positing essentialism may be a useful heuristic for everyday induction about the biological realm, but it may not be useful for scientific biology (Peter Godfrey-Smith 2009, 11ff).
[22] Reliance on cultural transmission also still requires an "origination" story for an idea, one that explains how the idea emerged so as to be transmitted, and the failure of the salience of visual differences account in children suggests its implausibility as an origination story.

children's knowledge of race seems to outrun their knowledge of the perceptible, physical differences prototypically associated with race. For instance, children misidentify their own racial group using dolls, but not using verbal labels. And children can use race as a category in free sorting tasks, and also as a basis for racial prejudice, but less so in the selection of playmates (see Hirschfeld 1996, e.g. 90-1, 138-9). Hirschfeld argues that this and other evidence suggests children do not acquire knowledge of races by inferring categories based on visual differences but instead come with "propensity to learn about a specific phenomenon: the world of human kinds" (64).[23] This propensity leads them to acquire and organize beliefs regarding human groups well in advance of their ability to coordinate them with perceptual learning or with action.

A different source of evidence that also suggests children's cognition of human groups precedes their visual-perceptive understanding of racial categories comes from Katherine Kinzler and colleagues' work on childhood preferences of persons based on accent. Preference for like-accented persons emerges early and robustly and outruns preference for like-raced persons (Kinzler et al. 2009). The evolutionary cognitive paradigm we've been considering offers a nice explanation for this: children may use accent as an indicator for important social groups, but only more mature children have learned to associate ideas about such groups with the bodily differences that are characteristic of racial difference in the contemporary U.S.[24]

One possible upshot of data from both Hirschfeld and Kinzler is that children seem to be on the lookout for specific sorts of information about important human groups from a very young age, and can even make some use of such information, but it takes them some time to come to believe that the visual differences sometimes considered paradigmatic of racial difference (as in U.S. culture) are markers for important social groups in contemporary America.[25]

It should be repeated that a simple empiricist hypothesis on which racial beliefs are extracted from salient visual differences is only one of many possible challenges to an evolutionary cognitive explanation, and it is quite a bit simpler than a more plausible hypothesis emphasizing an important role for transmitted culture

---

[23] This sort of account is echoed in Doug Jones' (2009) work on Brazilians' concepts of race that we discuss in Section 3.2. Jones concludes that, "The *conceptual* structure associated with racial terms may differ from their *perceptual* structure" (258).

[24] An early preference for using differences in accent over appearance may be evidence of the adaptation of the mechanism underlying folk racialism for thinking about social groups (if differences in accent but not bodily features would have been common over the course of human evolution). This fits well with Gil-White's (2001a) hypothesis.

[25] Other data from Hirschfeld show a preference for using properties like skin color in inferences about people but not artifacts (see, e.g., Hirschfeld 1996, chapter 4).

that is at the heart of most representational constructionism. Ruling out this simple view does not show that a more sophisticated form of representational constructionism cannot explain the phenomena. It simply raises the bar by suggesting the theoretical sophistication with which even young children are approaching the cognition of human groups.

3 Essentialism across Cultures

We have reviewed some data, much of it developmental, that offers evidence in favor of the evolutionary cognitive hypothesis that essentialist elements of racial thought are caused by an innate, domain-specific mechanism. However, other explanations of the developmental data remain possible, including ones that are consistent with the HERE hypothesis. Consider, for example, the possibility that children might acquire a culturally transmitted theory: a complex biological or racial understanding from the rich cultural traditions in which they are immersed from birth. The evidence for the early emergence of essentialist thinking in the biological domain may make the acquisition of such a theory seem implausible, but this is hardly decisive. Perhaps we simply have not figured out how the acquisition occurs. Or consider the possibility that while children have an innate predisposition to essentialize *some* things, say, biological kinds, they nonetheless inherit a culturally transmitted essentialist-construal of human groups.[26]

Both hypotheses point to shortcomings of considering only developmental data. Certainly, to have much faith in the human species-typicality of a cognitive mechanism or predisposition, we need to look more broadly than developmental psychological data from mostly American children (Henrich et al. 2010).[27]

In this section, I consider cross-cultural evidence of lineage essentialism specifically about human kinds. First, I briefly review some qualitative historical evidence for essentialism about human groups in Classical Greece and Rome (Isaac 2004), as well as in Chinese (Dikötter 1992) and Indian (Bayly 1999) thought. And then, I turn to focus especially upon two apparently hard cases for the evolutionary cognitive theorist: Brazil and the Vezo of Madagascar. These cases are hard cases because of the persuasive evidence that race-like categories work differently in these cultural milieus, *even with respect to the features that evolutionary cognitive theorists regard as constrained.* Nonetheless, a careful review of the experimental evidence from these cases suggests that a disposition

---

[26] This is consistent with Smedley and Smedley's view of a conceptual break on which older parts might have been assembled into a new way of thinking about human groups (2012, 25).

[27] Cf. Gossett 1963, chapter 1, for a range of additional cases that suggest an essentialist interpretation.

towards lineage essentialism is present though its manifestation can be altered by cultural context.

3.1 Cross-cultural qualitative evidence of lineage essentialist thinking

3.1.1 PREMODERN EUROPE AND THE MEDITERRANEAN

Consider first that, despite evidence that color prejudice was not especially prevalent among the civilizations of the ancient Mediterranean (Snowden 1983), ancient Greeks and Romans did label specific human groups, hypothesize about their distinctive national characters, and also hypothesize about the origins of those differences (Isaac 2004; Kamtekar 2002). Crucially, even where theorists endorsed quite different accounts of the origins of human difference, they recognized that these differences were preserved in inheritance from parents to children over many generations (Isaac 2004, 74). Medieval Europeans also recognized distinctive, reproducing human groups, and they debated whether racial differences were evidence of multiple creations (Fredrickson 2002, chapter 1).

3.1.2 CHINA

There is a prominent tradition of classic Chinese thinking manifest in Confucian and Mencian texts that emphasize the unity of humankind, stressing that peoples of the world were one family or "all under heaven" (*tianxia*). But ideas that look considerably more like Western racialism emerge at a number of other places in Chinese history. Frank Dikötter highlights the way in which race-like ideas played a historic role, for example, in a nationalist response to the emergence of Buddhism in China (1992, 18ff). For instance, the astronomer and mathematician He Chengtian wrote:

The inborn nature of Chinese is pure and harmonious, in accordance with altruism and holding to righteousness... Those people of foreign countries are endowed with a hard and obstinate nature, full of evil desires, hatred and violence.   (Cited in Dikötter 1992, 19)

Much later, during the Ming Dynasty (1368–1644 CE), Ming-Mongol conflict was often cast in racialist terms:

The discussions of strategy are striking in their lack of Confucian-Mencian references to the Han and barbarian being of one family or similar in essence. Instead the Mongols were almost uniformally characterized as subhuman "dogs and sheep," "fierce and wild" opportunists with insatiable appetites for the wealth of Han China. As the Hong Wu emperor stated, the border tribes "are not of our race; their hearts and minds are different" (cited in Yang 1990: 284). In a text on statecraft that he is said to have compiled and annotated personally, the Yong Le emperor cited a Han dynasty assessment, "As for the Rong and the Di, in all respects they have a different essence [from the Han]. There is no difference between them and birds and beasts" (Ming Chegzu, 1410, 524).   (Johnston 1995, 187)

All this suggests that historic Chinese ideas about race are more diverse—and in unfortunate ways more familiar—than a reading of Confucian and Mencian texts would suggest.

3.1.3 INDIA

Indian Castes have often been compared to race despite the remarkable and apparently distinctive features of castes (for example, the distinct *jati* and *varna* caste systems, or the relationship of castes to ideas of purity and contamination). Still, while quite different, castes are believed to have distinctive characteristic properties that are reproduced in the children of caste members. The historian Susan Bayly writes:

> both in the past and today, those sharing a common caste identity may subscribe to at least a notional tradition of common descent, as well as a claim of common geographical origin, and a particular occupational ideal.... In theory... the central characteristic of 'caste society' has been for many centuries the hierarchical ranking of castes or birth groups. The implication here is that to be of high or low caste is a matter of innate quality or essence. This is what is said in many scriptural codifications of caste ideals; in real life, these principles have often been widely contested and modified. Nevertheless, even people who came to reject caste principles either recently or in the more distant past are at least likely to have been familiar with these notions of corporate moral essences or qualities, meaning that in 'caste society', gradations of rank and precedence are innate, universal and collective. The implication would be that all who are born into so-called clean castes will rank as high, pure, or auspicious in relation to those of unclean or 'untouchable' birth, regardless of wealth, achievement or other individual circumstances. (Bayly 1999, 10)

Though castes have their own cultural history, the lineage essentialist assumptions that apparently structure them look quite familiar.

3.1.4 LINEAGE ESSENTIALISM REAPPEARS

These different systems of human groupings are all quite different from one another and also from contemporary racial groupings, but crucially, they all seem to involve groups of persons that are presumed to share unseen properties that explain differences and are transmitted to their children, and they all plausibly reflect cognitive propensities to apply principles to human groups that also lie behind folk reasoning about the biological domain.

All this suggests that lineage essentialism about human groups exists outside the modern European-American context, a conclusion reinforced by other experimental work, including work by Francisco Gil-White (2001a) on patterns of lineage essentialism in Mongolia as well as Martin Kanovsky's (2007) exploration of ethnicity and essentialism in the Ukraine.[28] If so, then the HERE hypothesis looks to be mistaken.

However, we might think that this is too easy on the evolutionary cognitive view. The evolutionary cognitive view predicts that the cognitive propensity for lineage essentialism will emerge robustly across cultures, but there are widely reported to be exceptions and amendments to such essentialism. In Sections 3.2 and 3.3, I consider two "hard" cases for the evolutionary cognitive theorist: Brazil and the Vezo of Madagascar. These are hard cases for the evolutionary psychological theorist because both cultural contexts seem to involve something like ethnic or racial group classifications that are not understood in a lineage essentialist manner, casting doubt on the evolutionary cognitive implication that a propensity for lineage essentialism is nonetheless present. I argue that a careful look at the experimental evidence suggests this doubt is not warranted. The propensity for lineage essentialism is present in these cases, but it is not always manifest.

3.2 Brazil

In much of Latin America, there are a range of skin colors and phenotypic differences that are similar to those in the United States. At the same time, it is often said that there is little or no racism, or that there is not true race consciousness or categorization, at least in contrast with the United States. Fredrickson observes the apparently greater historical fluidity of Latin American group categories:

categories [of white, *mestizo*, and Indian] lacked the rigidity of true racial divisions, because aspirants to higher status who possessed certain cultural and economic qualifications could often transcend them.   (2002, 39–40)

Accounts of fundamental differences between racial understandings and experiences in Latin America are, in fact, so widespread that it would be foolish to deny them, but, for present purposes, our question is: do the sorts of differences we find between, for example, Cuba or Brazil and the U.S. consist in different assumptions of lineage essentialism? Or do they stem instead from some other differences? In addressing this question, I consider especially the example of Brazil.

Numerous observers have commented on the differences among Brazilian and American racial systems (e.g. Robinson 1999; Sheriff 2001). The Brazilian sociologist Oracy Nogueira suggested that the distinction amounted to a difference

---

[28] Though to be sure, Kanovksy emphasizes that lineage essentialism is only one sort of essentialized identity. Interestingly, however, his way of showing this involves using language as a marker of ethnicity, which may simply pit one form of lineage essentialism against another.

over the role that one's origin or lineage played in classifying a person, a suggestion subsequently taken up in cross-cultural psychological work by Marvin Harris and Conrad Kottack.[29] The idea is that relevant social groups in Brazil are marked by appearance (i.e. by skin color, hair type, and so forth), but that this sort of identification is not tied to one's parents' group membership. Rather, appearance is a defeasible visible marker for important social categories (e.g. for economic status). In effect, this hypothesis is that Brazilian social categories group the same phenotypic range that is present in the United States in a way that makes little or no assumption of lineage essentialism.

Harris and Kottack explored this suggestion in Arembepe, Brazil. They employed the simple task of showing a photograph of three sisters of differing skin color, and eliciting from 100 neighbors judgments about the group membership of the three sisters. What they found was that in only six of 100 cases were the three sisters all identified as belonging to the same group! From this (and follow-up investigations) they concluded that Nogueira was right and that descent lacks the role in Brazilian social categorization that it plays in American racial categories.

Further work by Harris (1970) used a related paradigm in which drawings of phenotypically diverse individuals were provided to Brazilian subjects. The range of descriptions that these drawings elicited turned out to be irregular. Harris concludes:

If there is an orderly principle by which *morenos* or *mulatos* are distinguished from *brancos*, *pretos*, *sarards*, *alvos*, *claros*, and *cabo verdes*, it is an extremely complex one. At the moment it seems as if Brazilians will call almost any combination of facial features by the terms *moreno* or *mulato* with a high but unpatterned frequency. (1970, 12)

Harris's data show remarkable diversity in the application of terms applied to the same individual—underscoring again the sense of fluidity surrounding Brazilian social categories and suggesting the absence of a central place for lineage in reasoning about category membership.

This work amounts to reasserting the HERE hypothesis as an explanation for the emergence of racial essentialism in the U.S. but not Brazil, and it has become central to what is now a common view about the character of Brazilian social categories. Surely, the thinking is, if lineage essentialism (or some bias for acquiring lineage essentialist representations) were innate, they would be common in Brazil—a country whose phenotypic range and cultural inheritance overlap heavily with Europe and the United States.

[29] Nogueira (1955), cited in Harris and Kottack (1963, 203–4).

The Brazilian example has led evolutionary cognitive theorists to responded in two ways: first, they deny that lineage essentialism is in fact absent in Brazilian social categorization (Gil-White 2001b; Jones 2009). Second, they deny that an evolutionary cognitive account of racial cognition predicts that Brazilian categories should operate in the same way as the United States (Jones 2009; Mallon 2010). Consider each.

1. Does the available evidence show that Brazilian categories are not lineage essentialist?

Gil-White (2001b) has questioned whether data of the sort Harris and his colleagues have gathered in fact support the view that Brazilian racial categories are very different from those of the U.S. He notes that by relying on visual appearance, Harris and colleagues may have led their subjects to interpret their categorization tasks as involving features that are merely indicative of category membership (e.g. skin color, hair color, and so forth) instead of forcing them to reason about features that they might be inclined to treat as criterial—e.g. as necessary or sufficient (for example, lineage) (2001b, 226ff; cf. Jones 2009). This problem was accentuated by other methodological problems, including asking subjects to categorize pictures using multiple different (and nonequivalent) sentences—e.g. "what race is this person" versus "what kind" or "what color" (in Harris 1970). Gil-White goes on to note that Harris's methodology also fails to allow for unmeasured variation in the naming of racial categories, ambiguity between labels as names of races and as labels of colors, and variation in the use of ordinate or superordinate racial categories.

The anthropologist Doug Jones (2009, 258ff) pursues Gil-White's critique further by directly questioning subjects in the region of Bahia, Brazil about the categorization of children of parents whose categorization is known. For example, Jones asked subjects "if the father is black and the mother is black" then "the child is?" Jones's study directly addresses the role of lineage in determining category membership. After subjects answered, Jones proceeded with a follow-up question, for example, if subjects answered that the child was black, Jones asked, "*could* the offspring in question be white?" The most striking initial feature of Jones's data is the nearly overwhelming agreement that when both parents are of a race, the child will be of that same race (over 80 per cent for each combination). In follow-up questioning, subjects revealed essentialist reasoning about mixed persons:

In explaining,—spontaneously or in response to prompting—how it was or was not possible for, say, two mixed parents to have a white child, or two black parents to have a mixed child, subjects (when their answers were forthcoming) referred repeatedly to nonvisible, internal traits and family background. (Jones 2009, 261)

And in follow-up questioning using classic "transformation" questions, subjects also held racial identity to be fixed under transformation of surface features (so, for example, they held that one cannot change one's race via plastic surgery). All this leads Jones to conclude that, when asked the right questions, "Brazilians... have an essentialist conception of race" on which it is "an inherited, underlying property of individuals" (2009, 264).

2. *Does an evolutionary cognitive explanation of racial essentialism predict that racial essentialism will be as important in Brazilian social classification as in U.S. racial classification?*

A prior question to ask is whether the evolutionary cognitive account really entails that Brazilian social categorizations should exhibit racial essentialism in the way North American racial categories do.

In fact, the suggestion that they should already contains Nogueira's and Harris's error, for the presupposition is that the relevant input for prescientific racial thinking is an array of local bodily phenotypes. Since this array is reasonably similar between North and Latin America, the idea seems to be that racial categories (were they cognitively rather than culturally determined) ought to be very similar as well. However, the assumption that our cognitive machinery takes *appearances* as the primary input in racial theorizing is itself an assumption that evolutionary cognitive theorists can and do reject. Recall, for instance, Hirschfeld's argument that children's theorizing of human groups outruns their ability to apply distinctions perceptually. Instead, categorization structured by the sort of lineage essentialism we have been focused upon here ought to pay special attention to parent kind memberships and offspring relationships.

If this is true, then where parental kind memberships or offspring relationships are different, we ought to expect the resulting system of social classifications to be different as well. Consider Fredrickson's claim that during colonial exploitation of Latin America by the Spanish and Portuguese:

Indians were brutally exploited by the possessors of *encomienda* and the proprietors of silver mines and haciendas, but the purity-of-blood doctrine was never systematically applied to those with part-Indian or even African ancestry. An attempt to order society on the basis of *castas* defined in terms of color and ethnicity eventually broke down because the extent and variety of *mestizaje* (interracial marriage and concubinage) created such an abundance of types that the system collapsed into the three basic categories of white, *mestizo*, and Indian. (39–40)

Fredrickson's claim is, in effect, that the ubiquity of intergroup breeding undermined the capacity of prescientific essentialists to sustain an essentialist classification regime, and so mixed-race categories were introduced and normalized (see, e.g., Telles 2002 for a similar view). This historical fact may be the central piece of what drives other apparent differences between North and Latin America (e.g. the comparative lack of importance of race in Brazil). And this is perfectly compatible with the evolutionary cognitive account of racial essentialism.

This explanation of difference relies on an empirical claim, namely that intergroup interbreeding was historically more ubiquitous in Latin America than in North America. Even allowing for this to be true, though, it remains unclear why interbreeding in North America, historically or contemporarily, has not (yet) resulted in a similar revolution in racial classification. Another part of the answer, it seems clear enough, is that U.S. practices and institutions of racial classification have long employed rules of hypodescent (like the "one drop of blood" rule) to classify descendants of black persons as black (e.g. Cohen 1949), while in Brazil a range of intermediate categories are recognized that may appear in the range and diversity of the racial labels elicited by investigators using pictures.[30] In short, Brazil and the United States vary not only in their racial history, but also in the ways they culturally enforce racial boundaries.

However, Doug Jones (2009) and I (Mallon 2010, 2013a) have separately argued that such cultural variation is to be expected on (some versions of) the evolutionary cognitive account. Recall the parallels between racial and biological cognition. One plausible explanation for these parallels is that whatever mechanism underlies racial cognition was initially adapted to thinking about (certain sorts of) biological kinds, and then perhaps later exapted for thinking about social kinds (e.g. Gil-White 2001a). Suppose that, in particular, the mechanism was originally adapted to thinking about members of fully or largely reproductively isolated biological populations (e.g. it's adapted to thinking about classic species). In contrast to the *proper domain* of this mechanism—i.e. the domain it was originally adapted to operate on (Sperber 1994)—humans races are not species.[31] In particular, although human populations sometimes exhibit partial reproductive isolation (see Andreasen 1998; Kitcher 1999; Mallon 2006 for discussion),

---

[30] Telles (2002) puts the point in a more general way, suggesting that in Brazil public laws centered on race (and thus creating legal classification) were rare (422).

[31] If this mechanism was subsequently adapted to the domain of human groups (i.e. if it is exapted), then the proper domain will have shifted (the new domain will also be a proper domain). However, the point made in text remains: the mechanism may still be structured by its historic function. It may be that the cognitive mechanism is not adapted to essentializing mixed-race persons because there has been no selective benefit to doing so.

human populations frequently and readily interbreed and have done so (Templeton 2013). Thus, this interpretation of the cognitive mechanism underlying racial cognition has it that the mechanism is originally adapted to populations that rarely interbreed, but it is being applied to populations that frequently interbreed. It should therefore be no surprise that such a mechanism fails to neatly draw conclusions about mixed-race persons. Jones and I have argued that this suggests a prediction of greater conventionality regarding classification of mixed-race persons than of nonmixed-race persons (Mallon 2010), with Jones arguing that this is in fact the case (Jones 2009, 2064ff).[32]

Evolutionary cognitive accounts suggest that we ought to expect strong overlap in classification judgments of nonmixed-race persons between the United States and Brazil, and this is just what we find. In both countries, a child of two parents of the same race is typically classified as a member of that race (when the parents' race is known). But endorsement of an evolutionary cognitive account gives us no special reason to expect one rather than another way of classifying mixed-race persons. And different cultural choices and practices, together with different histories, can mean very different social categorizations result. Acknowledging this fact does not mean insisting that Brazilians lack lineage essentialist racial categories so much as it suggests that, in the Brazilian context, such categories fail to be of much use in categorizing persons.

3.3 Vezo

While the Brazilian experimental evidence is powerful, it remains consistent with the idea that racial essentialism was introduced in modern Europe and America and somehow imported by Brazil. However, other evidence points to the same conclusion. Perhaps most interesting is Rita Astuti and colleagues' research among the Vezo of Madagascar (Astuti et al. 2004).

The case of the Vezo is especially interesting, for as Astuti's earlier fieldwork demonstrated, Vezo adults view their identity as not a matter of biological inheritance. Astuti writes, "The term 'Vezo' also denotes a people. The Vezo often point out that their name means 'paddle,' a name which indicates who they are, people who struggle with the sea and live on the coast." (1995, 1). Astuti goes on to relate that,

---

[32] Discussion of mixed-race persons has long been part of the social philosophy of race, in part due to the important work of Naomi Zack (e.g. 1993). The hypothesis here dovetails nicely with Zack's (1993) idea that attention to cases of mixed-race persons undermines racial thinking since mixed-race cases may dumbfound the mechanism by which we think about race, leaving us falling back on convention or stipulation for determinate categorization judgments.

Any attempt on my part to learn new words related to fishing or sailing, for example, prompted my instructors to explain that all the people who fish and sail are Vezo. Similarly, when I showed a group of young men a map of the coastal region, they told me that all the people who live along the coast, near the sea, are Vezo. (2)

Because Vezo identity is fixed by occupation rather than by descent, one can acquire or lose it. As such, Vezo adults' understanding of Vezo identity is genuinely different from the lineage essentialist categories of ethnicity and race that we have been considering.

However, experiments by Astuti and colleagues (2004) suggest a quite interesting developmental pattern. When given an "adopted at birth task" in which Vezo children are raised by Masikoro (an inland group) or Karany (descendants of immigrants to Madagascar from the Indian subcontinent), adult Vezo consistently judge that group identity is fixed by way of life (2004, 44ff., study 1), fitting with Astuti's earlier ethnographic work on Vezo conceptions of Vezo identity. However, six–thirteen-year-old children show an *adopted parent bias* (the child will have the group identity of the adopting parents) when the switch is Vezo-Masikoro, but a *birth parent bias* when the switch is Vezo-Karany. Astuti et al. write: "Thus, as a group, children took Vezo identity (relative to Masikoro) to be determined by where one grows up, but Vezo identity (relative to Karany) to be determined by one's birth origins" (Astuti et al. 2004, 64, study 2). This is crucial, since Vezo-Masikoro identity is a topic on which children will have had more specific instruction, while the Vezo-Karany question is novel and requires inferences based on background knowledge. What this suggests is that Vezo children reason about the novel Vezo-Karany task using the sort of implicit folk theory suggested by evolutionary cognitive theorists, but that this theory is supplanted by explicit instruction in the Vezo-Masikoro task.[33]

We already saw (in our discussion of Brazil) that cultural variation in racial theories is compatible with the evolutionary cognitive account. However, Astuti's evidence suggests intercultural variation even in the core of what evolutionary cognitive theorists say is explained: the essentializing of group identity. At the same time, this evidence shows that the fact of cultural variation by itself does not undermine the evolutionary cognitive thesis that the mind is evolved with a predisposition to essentialize such identities, for this predisposition seems evident in Vezo children, but can be transformed or overridden by local culture. In fact, the data she has gathered are (admittedly

---

[33] Kanovsky (2007, 250ff) offers a reading of Astuti (2001) that prefigures the one here.

prima facie) evidence against both simple culturally transmitted theories and culturally transmitted transfers of essentialist construals (for both would predict that the absence of essentialism about human groups is the default).

3.4 Innate, domain-specific, and species-typical

Both the qualitative and experimental evidence suggests that lineage essentialist reasoning about human groups is considerably more widespread than the HERE hypothesis would suggest. While it remains possible to argue that multiple traditions invented essentialism independently, or again that the apparent existence of essentialist thinking in other traditions (e.g. Madagascar or Mongolia or Brazil) actually is the result of cultural influence from Europe or North America, I argue that the cross-cultural work when placed alongside the developmental evidence supports a different hypothesis, namely that the evolutionary cognitive theorists are correct and that essentialist thinking about human groups is the product of a somewhat culturally canalized, domain-specific, and species-typical mechanism that manifests itself quite broadly.

I say "quite broadly" rather than universally, for as we have seen, culture and social environment can and do check this manifestation. As we noted in Section 1.4, it does so among contemporary biological theorists, who widely regard essentialism as false for both species and races. It also does so in the Vezo context, where essentialist assumptions of children seem to be overridden via explicit instruction. And it does so in the Brazilian context, where a different cultural practice (one emphasizing mixed-race rather than hypodescent means of classification) and a different social environment (one in which mixed-race persons are very common) undermine the easy applicability of lineage essentialism.

In fact, there is further evidence that suggests some contingency of lineage-essentialist assumptions. For example, a range of evidence from within cultures shows that cultural position affects endorsement of lineage essentialism. For instance, in Israel, religious Jewish children essentialized labeled groups more readily than less religious children (Birnbaum et al. 2010, study 1). And while American subjects hold that the behavior of a brain transplant recipient is determined by the ethnicity of the brain donor, Indian Brahmins do not (Mahalingam and Rodriguez 2006). To put it another way, a wide range of factors determines what individuals and groups believe and do. What we can conclude from the present evidence is that influence by the psychological essentialist mechanisms posited by some evolutionary cognitive theorists is likely one of them.

4 Critiquing of "Folk Essentialism" and the Conceptual Break Hypothesis

Thus far, I have presented a defense of what I take to be the central strand of the research on "folk essentialism," focusing on its implications for race and the Conceptual Break Hypothesis, and I think this reading is both plausible and grounded in the data. However, I have elided a number of critiques of work on folk essentialism to which I now turn. I begin with *deflationary* or minimal readings of the inferential patterns explained by folk innateness. I then turn to consider *splitting* critiques of folk essentialism that suggest that the various features of folk essentialism are dissociable into independently varying components. And I finally return to concerns that folk essentialism is *culturally particular*.

4.1 Deflationary or minimal readings of the data for folk essentialism

Members of the evolutionary cognitive research tradition that I have been considering posit folk essentialism on the grounds that it offers the best explanation of the available data. However, there are others, even those perhaps broadly sympathetic to an evolutionary cognitive research agenda, who are skeptical of this attribution.

Michael Strevens (2000) has provocatively claimed that the evidence in favor of psychological essentialism can all be accounted for by a different, *minimal* hypothesis: viz. that adults and children alike perform the sorts of categorizations and inferences we have been reviewing because they endorse certain laws of nature that connect membership in a kind with certain kind-typical properties (e.g. "All tigers have stripes"). Strevens makes a powerful case that many of the results that cognitive psychologists have suggested are to be explained by appeal to prescientific beliefs in "underlying natures" can in fact be explained more simply by appeal to folk beliefs in such laws. As he points out, it seems possible to think that, for example, there are laws connecting kind members with particular traits, while being neutral about exactly what (an essence, or something else) it is in virtue of which kind members participate in such laws.

Strevens's argument is controversial (see Ahn et al. 2001; Strevens 2001a, 2001b), and I do not believe it. However, if it is correct—or if some *other* alternative minimal explanation of the categorization and inference data is correct—then this might seem to pose a problem for the argument I am making here since it suggests the folk do not really make assumptions about hidden essences.[34]

---

[34] See also Fodor (1998, 154–5), who expresses skepticism that the psychological work on folk essentialism reveals a real commitment to belief in essence and reveals a folk understanding of

But whatever the merits of these alternative explanations, they can be safely ignored here. My interest is in insisting upon the relative invariance across cultures of *whatever mechanisms* (essentialist or minimal) that do the explaining, but I do not need it to be the case that the mechanisms be psychologically essentialist.

We can see this by noting that if a psychologically nonessentialist explanation of the prescientific practices of categorization and inference studied by evolutionary cognitive psychologists is correct, then it is plausible that a similar, psychologically nonessentialist explanation of the prescientific practices of categorization and inference studied by social constructionist defenders of the Conceptual Break Hypothesis is also correct. The result would be that both groups would have to recast their explanatory hypotheses in a psychologically nonessentialist way. But the central question—"Are the mechanisms that give rise to the patterns of judgments and inferences identified by both constructionist and evolutionary cognitive researchers an invention of the European-American cultural tradition? Or are they features of an innate, species-typical, and domain-specific cognitive architecture?"—would remain. And the evidence in favor of an evolutionary cognitive account of essentialism would also favor the cultural canalization of a nonessentialist alternative explanatory mechanism.

4.2 Splitting essentialism

Our distinction between broad essentialism and lineage essentialism suggests another possibility: that what we have characterized here as folk essentialism is in fact a range of distinct cognitive proclivities that may only contingently co-occur in the case of racial essentialism. Evidence for this takes a number of forms. Some researchers have shown that several elements that co-travel in the empirical literature as "essentialism" can be distinguished and shown to vary somewhat independently within or across cultures (e.g. Haslam et al. 2000; Moya and Boyd 2015). But for our purposes, we can allow that the cognitive capacities or dispositions underlying folk essentialism can be "split" into sub-capacities or dispositions without worrying that they do not exist, and experimental demonstrations showing that such sub-capacities or dispositions do not always co-occur does not show that they do not reliably or commonly co-occur.

---

natural kind concepts (which he takes to be a late and sophisticated achievement). Machery et al. (ms) follow Fodor in this skepticism, providing what they take to be cross-cultural evidence against the cultural invariance of psychological essentialism.

4.3 Essentialism as culturally particular

A number of recent researchers have combined the "splitting" of essentialist judgments into components with experimental and cross-cultural work in order to show complexity and variation in essentialist judgments within and across cultures (Haslam et al. 2000; Moya and Boyd 2015; Olivola and Machery 2014; Machery et al. ms). Indeed, we have already seen evidence of cultural variation among group representations; even in precisely those features that the evolutionary cognitive program seems best equipped to explain. For example, we have seen that:

- Race-like representations may vary in regard to their role in producing or ameliorating intergroup hierarchy or conflict, as in cases of Chinese empire, Indian caste, and Brazilian colonialism.
- Race-like representations may vary in their applicability in a context, as in context of ubiquitous "mixed" descent, especially in a context that does not employ a cultural convention of hypodescent.
- Race-like representations may be nonessentialist as the result of explicit instruction (as among the Vezo of Madagascar, or among modern, biologically sophisticated thinkers).

The psychological essentialist strategy has been to appeal to a disposition that may or may not be manifested in a particular cultural context, and this strategy can be repeated in the face of emerging evidence as well. The HERE hypothesis suggests that racial essentialism is a culturally local invention, perhaps of the nineteenth century. The existence of evidence for the manifestation of the disposition to essentialize human groups across cultures, history, and developmental ages suggests the HERE hypothesis is false, even if there are also other (perhaps many other) cases in which racial essentialism is not manifest.

This is not a complete explanatory victory for innate, species-typical explanation. If we ask the broad question:

Why are European and American racial representations essentialist rather than not?

The evolutionary cognitive theorist can plausibly answer:

Because of innate, domain-specific, and species-typical features of our cognitive architecture.

In insisting that a trait is innate and species-typical, evolutionary cognitive theorists are insisting on a kind of developmental regularity. But this regularity can be and is (sometimes, perhaps often) disrupted.

But it also follows from our argument that the social constructionist about representations can answer our broad question with:

Because of the histories, cultural practices, and beliefs of the communities in which these representations figure.

Racial representations emerge in a complex causal field with many determinants that could, in principle, be manipulated or intervened upon to change the outcome. In this explanatory context, both constructionist and evolutionary cognitive mechanisms can play a role as part of a complete causal story even as the evolutionary cognitive story undermines claims of cultural invention or uniqueness.

5 The End of the HERE Hypothesis and Whither the Conceptual Break Hypothesis?

Some defenders of the Conceptual Break Hypothesis rely on the HERE hypothesis, but a range of evidence supports the evolutionary cognitive proposal that racial essentialism is a product of an innate, domain-specific, and species-typical mechanism that disposes minds to racial essentialism across cultural contexts. This suggests that the HERE hypothesis is mistaken. With it goes a likely route for vindicating the Conceptual Break Hypothesis.

Like social constructionist explanations, the evolutionary cognitive explanations I have defended here are, prima facie, *debunking*: they undermine realist explanations of essentialist representations that explain their essentialist content by appeal to their being causally and rationally constrained by biological reality. In this respect, social constructionist and evolutionary cognitive explanations of racial representations find common cause against realist explanations of racial essentialism, a fact that allows social constructionists about other aspects of racial representations to avail themselves of evolutionary cognitive explanations of racial essentialism without abandoning skepticism about racial essences.

If evidence suggests that the HERE hypothesis is mistaken, are there other possibilities for defending the Conceptual Break Hypothesis? Defenders, it seems, have two options:

1. They could offer some other explanation of the content of the conceptual break, and then show that this content emerges only locally in the modern European-American context.
2. They could abandon the Conceptual Break Hypothesis all together, and explain modern cultural shifts surrounding race in ways that do not depend on a specific, significant shift in the content of people's beliefs about human groups.

My own sympathies lie with the second strategy. In this, I join with a great deal of constructionist academic work on many human categories suggesting that the modern ways of conceiving and constituting human groups involve some sort of break, but I deny the Conceptual Break Hypothesis as an account of it—deny that the break involved some significant shift in how people thought about the human groups that we now think of as races. It instead involved some other sort of shift, perhaps, a political, institutional, or technological shift.[35]

One way to develop such a view is to suggest that what emerged for the first time in the nineteenth century were racial theories that became widely shared, effectively displacing a mishmash of essentialist racial theories with a more unified essentialist account of racial difference and hierarchy that could serve as a transnational rationalization for European colonial ventures and American slavery. It is easy to think of the products of scientific theorizing as theories distinguished by their novel contents. But scientific theorizing has many products, and included among them is sometimes widespread consensus that can (because of its capacity to organize and rationalize behavior, institutions, and policies on a wide scale) effect repercussions that a hodgepodge of more local theories cannot match.

But a qualification of this endorsement of the second strategy is in order. Many philosophers view the content of concepts to be at least partially externally fixed by their referents (e.g. Putnam 1975; Burge 1979; Fodor 1994). I argue in what follows that human category terms may come to refer to human kinds that are themselves constituted, in part, by human representational practices. It follows that if a set of practices (say those surrounding representations of *white* or of *black*) first emerged in the modern era, then any kinds they constitute would as well. On the externalist assumption that a kind constitutes (at least partially) the meaning of the representations that refer to it, a change in the meaning of those representations will have occurred as well. Success in the second strategy thus makes possible success in the first, albeit a different kind of success than Conceptual Break defenders usually conceive. To carry it off, we need to develop a constructionist account not of our representations but of the categories themselves.

---

[35] Something like this view sometimes appears right alongside claims of a conceptual break. For example, Fredrickson emphasizes the role that representations of race play in justifying or giving "legitimacy" to a "racial order" (2002, 24); and Paul Taylor talks about modern racialism being "self-conscious" in that "people explicitly appeal to race in organizing the social world and their perceptions of it" (2013, 22).

2

Constructing Categories
Concepts, Actions, and Social Roles

The most provocative social constructionist claims are those that concern the construction, invention, creation, or sustenance of nonrepresentational features of the world, including human categories understood ontically, as things in the world. Thus, one common category constructionist claim goes beyond the assertion qualified in Chapter 1 that racial representations are constructions to the claim that race itself is an invention. We saw this in Stuart Hall's claim in the Introduction, and it is easy to find elsewhere. For instance, the philosopher Paul Taylor writes that,

> White supremacist societies created the Races they thought they were discovering, and the ongoing political developments in these societies continued to re-create them...All of this is to say: our Western races are social constructs. They are things that we humans create in the transactions that define social life. (2013, 179)

Similarly, Michael Omi and Howard Winant write that:

> The effort must be made to understand race as an unstable and "decentered" complex of social meanings constantly being transformed by political struggle...the concept of race continues to play a fundamental role in structuring and representing the social world. The task for theory is to explain this situation. It is to avoid both the utopian framework which sees race as an illusion we can somehow "get beyond," and also the essentialist formulation which sees race as something objective and fixed, a biological datum. Thus we should think of race as an element of social structure. (1994, 55)

This core idea, that races are products of human sociocultural activity, is widespread among humanistic racial theorists.

Claims of the construction of gender categories are also ubiquitous. We mentioned Catherine MacKinnon's work in the Introduction, and to this we can add Judith Butler's provocative claim that gender is a kind of performance:

> That the gendered body is performative suggests that it has no ontological status apart from the various acts which constitute its reality...In other words, acts and gestures,

articulated and enacted desires create the illusion of an interior and organizing gender core. (1990, 136)

The suggestion is, again, that gender differences are, at least in substantial part, a product of specific social-cultural practices that could be changed.

Perhaps even more widespread than social constructionist claims about race or gender are social constructionist claims about homosexuality. In the Introduction and Chapter 1, we also noted Michel Foucault's influential claim that as ideas or concepts of sexuality shifted in nineteenth-century Europe, they became more committed to the idea that sexual behavior was indicative of importantly different, underlying kinds of people, actually producing categories of person. Others have followed Foucault with discussion of other diagnostic categories. Arnold Davidson (2001), for instance, writes of "perversion":

Perversion was not a disease that lurked about in nature, waiting for a psychiatrist with especially acute powers of observation to discover it hiding everywhere. It was a disease created by a new (functional) understanding of disease, a conceptual shift, a shift in reasoning, that made it possible to interpret various types of activity in medicopsychiatric terms. There was no natural morbid entity to be discovered until clinical psychiatric practice invented one. (2001, 24)

Similar constructionist claims about categories of person have shaped a wide range of discussions including constructionist work on emotions and antipsychiatric views of mental illness (e.g. Averill 1980a, 1980b, 1994; Szasz 1974; Scheff 1974, 1984; Goffman 1970).

Category constructionists hold that the creation or persistence of a category or its features is brought about, in part, by practices of representing humans as instances of the category, or of representing the category itself. More precisely, we can say that category constructionists hold that human mental states, decisions, culture, or social practices that represent a category C causally or constitutively explain how C came to have instances, how C continues to have instances, or how members of a category came to have or continue to have their category-typical properties.[1] Covert constructionist explanations of human categories contrast with explanations on which the existence, persistence, or features of instances (or putative instances) of C are explained by natural or biological kinds. Perhaps most commonly, the salient contrasting natural explanations advert to *biobehavioral kinds*: kinds that connect membership in a biological category with behaviors or dispositions to behave in category-typical ways.

---

[1] Because, e.g., *believes that p* is itself a human category, representational construction can be cast as a special case of category construction. I treat it as distinct for clarity regarding the special issues it raises.

But on their face, category constructionist explanations face serious challenges. It is not in general the case that representing a category or something as belonging to a category causes the category to be, or to continue to be, instantiated. Nor does it generally change the properties of a person to those represented of a category. Constructionists therefore owe explicit accounts or models of how such category construction is achieved, and specifically of the role that representing a category plays in its construction.

Satisfying this demand is all the more pressing because of the causal significance of these categories. Constructionists are apparently offering competing explanations for some of the most explanatorily powerful categories in the human sciences.[2] For instance, social and medical scientists tell us that race is correlated with residential patterns as well as differential health, wealth, educational, and economic outcomes. Similar arguments can be made for sexual categories, for gender, for homosexuality, and for a range of other candidate constructions. Whatever their source, the social and medical sciences often treat these categories as perfectly real and causally powerful, and reference to them figures in apparently powerful explanations. So, the constructionist owes an explanation not only of how categories can be constructed, but how they can be constructed so as to be causally significant.

Notice that one constructionist opponent, the biobehavioral theorist, has a clear explanatory strategy for addressing the causal significance of these human categories: appeal to natural or biological kinds. According to the biobehavioral theorist, category members are distinguished by the instantiation of biological natural kinds that underwrite the category boundaries and explain the typical features of category members. Taking this as our model, a comparative constructionist strategy suggests itself: we simply appeal to constructed (rather than natural) kinds to explain the causal significance of human categories, and we explain how these kinds are produced and sustained by appeal to categorical representations and the practices and other things that these representations structure. But what is wanted is to further specify how practices of representing categories and classifying persons as members of them come to construct those categories as real and as causally significant. It is this project I take up in this chapter and Chapter 3.

In this chapter, I consider three different accounts of the construction or invention of categories. The first *semantic individuation* account rests on a semantic strategy for individuating the human categories *thickly*, such that they are almost invariably local and (at least in part) culturally constituted. While this

---

[2] Root 2000; Mallon 2003; Alcoff 2006; and Mallon and Kelly 2012 offer similar framings of this problem.

is indeed a source of apparent conflict between constructionists and their opponents, I argue that we ought to understand some category constructionists as attempting to specify a kind or mechanism that offers a competing explanation of the differential features of a category that are of interest to us, offering an explanatory alternative to biobehavioral kinds.

In Section 2, I turn to consider Ian Hacking's analysis of category creation, one that understands construction as crucially involving specific concepts or meaningful descriptions in intentional actions. This *action analysis* does suggest a competing explanation of provocative claims of category construction and also the causal differentiation of category members, but I argue that such an account remains seriously incomplete.

In Section 3, I sketch the social-conceptual core of my own positive account, one based on *social roles* that are created when human category representations become common knowledge in a community. But in Section 4, I acknowledge that the mere existence of a social role does not by itself ensure that the social role will result in a causally relevant kind. I turn in Chapter 3 to suggest mechanisms by which social roles become causally significant kinds.

1 Individuating Kinds and Competing Explanations

A category will be culturally or historically local whenever the necessary conditions on belonging to the category involve culturally or historically local elements. Instantiating the institutional kinds *King* and *Queen of Carnival*, whose members are selected by the Krewe of Rex in New Orleans each year, necessarily requires satisfying a set of other culturally and historically specific norms whose emergence can be dated to the nineteenth century. But how do we know what the necessary conditions for falling under a category are?

One venerable way of understanding the necessary conditions for membership in a category is to understand these conditions as stemming from conditions on the application of the *concept* or *meaning* associated with the label of the category, where these can be found by understanding the description, conception, or theory associated with the label (e.g. Jackson 1998). As I noted in Chapter 1, it is difficult to determine what elements of an associated description of a term count as meaning constitutive. However, in general, opulent strategies that incorporate lots of beliefs about (culturally, spatially, or temporally) local elements into the definition of a category will result in categories that are more local while more austere strategies that incorporate fewer elements, and especially fewer culturally local ones, will result in kinds that are more general.

The anthropologist Clifford Geertz influentially recommended that anthropologists should pursue "thick descriptions" of phenomena, incorporating a great sweep of local conditions and understandings into those descriptions (Geertz 1973). In much the same spirit, some constructionist accounts of emotions have sought to specify the very specific cultural circumstances and understandings surrounding emotional behavior in different human cultures (e.g. Lutz 1988). Where we take these elaborate descriptions or specifications to serve as opulent definitions of the relevant kinds, then the kinds themselves almost invariably turn out to be local and culturally constituted (providing they are instantiated at all). For instance, if we define the meaning of "anger" as an appropriate feeling when:

the American flag is desecrated;
someone cuts you off in line at the store;
another driver gives you "the finger";
a careless person knocks over your Harley;

then it follows that "anger" will only apply in highly circumscribed cultural situations in which other practices, norms, and artifacts are also in place. Anger, on this understanding, will be a highly culturally specific category.

Similarly, if, in order to count as a racial concept, a concept must make specific use of the distinction between darker and lighter skin as an identifying factor, then many concepts—even those that look to involve assumptions of underlying, heritable essential differences among human groups—will not count as racial concepts. If, even more specifically, racial concepts are seen as labeling distinctions necessarily fixed by events in American history—the tradition of chattel slavery or the Civil Rights movement, for instance—then these concepts will have even more limited application to non-American places and times.

Elsewhere, Stephen Stich and I have suggested these differing strategies of fixing the meaning or reference of category terms sometimes drive disagreement in the social sciences over the cultural universality of human emotions and other traits (Mallon and Stich 2000). This is because in contrast to the opulent, culturally laden definitions of human categories favored in some humanistic social sciences, others pursue austere, thin ones. Evolutionary-cognitive theorists of the emotions, for instance, might understand emotion kinds as manifestations of innate suites of species-typical responses to certain schematic "core relational themes." Anger can then be understood as a certain sort of nervous system response that co-occurs with a certain phenomenology and a suite of facial and other bodily reactions that are activated in response to a perceived offense (Lazarus 1994). Austerely construed, "anger" can apply much more broadly across cultures.

To repeat, opulent construals of the meaning of terms can guarantee that provocative claims of cultural locality will turn out to be true, and austere construals produce greater generality.[3] These are differences in strategies for construing the meanings of terms, and there is no consensus on how to resolve them. The crucial point for Stich and I was that because such disputes are not empirically substantive, they are best seen as targets for clarification and stipulation before being set to the side (cf. Mallon 2006). Similarly, here I avoid arguments that depend upon (either opulent or austere) a priori stipulations of meanings for human category terms. Instead, I focus on the more substantive thought that some category constructionists aim to offer alternative, competing explanations of some of the *very same phenomena* that the biobehavioral theorist attempts to explain. In such cases, the disagreement is not merely verbal and can be empirically substantive. It is in these cases that greater clarity regarding the exact character of a socially constructed category would make the constructionist case more vivid and plausible.

2 Making up People and Intentional Action

Hacking has suggested that we can understand constructionist claims about the creation or invention of human kinds—what he calls "making up people"—by appeal to the requirements for intentional action (1986, 1995a). The basic idea is to explain how it is that members of different categories act differently by appeal to the differing content of the descriptions under which they act. While Hacking's own considered view is that an exclusive focus on intentional action is insufficient to explain category construction (1995b, 2007), clear consideration of his action analysis will allow us to see both advantages and shortcomings of a focus upon intention-driven behavior.

2.1 Necessary description category construction

Category constructionists hold that the construction of a category is brought about, in part, by practices of representing humans as instances of the category or simply of representing the category. This, in turn, requires a concept or conception *of* the human category, a way for thoughts or words to *mean* or be *about* that category. Thus construed, category constructionists are committed to the idea that in order for a constructed category to have instances, someone must conceive of it.

---

[3] The limit case of austerity could involve adopting a causal-historical account of the meaning and reference of a kind term on which *none* of the beliefs associated by speakers with the referent need be true of the referent (Putnam 1975; Kripke 1980).

The necessity of certain beliefs or descriptions or conceptions for category construction offers an appealing explanation for category invention or creation. Where other conditions of category construction are in place, and a novel way of conception or description of a category emerges, then the category itself emerges in tandem, "invented" or "created" alongside the practice of conceiving it. Such an analysis also makes sense of the category constructionist idea that changes to the beliefs or theories people have about a category are a route to changing the category itself—that representations control categories.

But we are still left with questions: what is it for a category to be constructed? And how, specifically, do category representations contribute to that?

2.2 Hacking, action, and identification

Hacking's action analysis is rooted in an account of intentional action. Hacking begins with G. E. M. Anscombe's (1957) influential view that human intentional action is individuated by the intentional description under which it occurs. A familiar illustration has it that raising one's hand in a meeting (as if to ask a question) can be the same bodily motion as raising one's hand on the street (as if to hail a taxi), or at a party (as if to attract the attention of a friend), but each of these motions are different *actions* in virtue of the intention with which they were undertaken. What Hacking puts in terms of descriptions, I will put in terms of the concepts that constitute the meaning of the descriptions. Thus transfigured, the intentional descriptions under which we act are individuated by their propositional content, where this is some function of the individual concepts that constitute that propositional content. So, for instance, if

Lucille intends that *she takes a nap on the couch.*

The propositional content is *she takes a nap on the couch*. And this proposition is constituted by combining concepts like *nap* and *couch*.

Hacking's idea (put in "concept" terms) is that we can explain the necessity of specific concepts for category construction by appeal to the necessity of specific concepts for specific intentional actions. Hacking points out that it follows from Anscombe's view that in order for a person to intentionally act under the description d, where d expresses the concepts $c_1, c_2, c_3 \ldots c_n$, the person must have the ability to intend d, and therefore must have or be able to express the concepts $c_1, c_2, c_3 \ldots c_n$. Lucille cannot intend to nap on the couch unless and until she has a description that can express *nap* and *couch*. Hacking explains:

Descriptions are embedded in our practices and lives. But if a description is not there, then intentional actions under that description cannot be there either: that, apparently, is a fact of logic. (1986, 230)

For Hacking, provocative instances of category construction emerge from the fact that no one can intentionally act under a description without having the concepts that constitute the meaning of the description. It follows that, for example, before the concept of homosexuality was available, no one could act as a homosexual. On Hacking's action analysis, intentional action under a category concept is essential to category construction. But why believe this? Why believe that constructing a category requires acting under a description whose conceptual content is about the category?

Hacking, and those like Davidson and Foucault that he is interpreting, is focused on cases in which an emerging label and concept of a category offers people a new description under which to act. He writes that:

When new descriptions become available, when they come into circulation, or even when they become the sorts of things that it is all right to say, to think, then there are new things to choose to do. When new intentions become open to me, because new descriptions, new concepts, become available to me, I live in a new world of opportunities. (1995a, 236)

Hacking has used this idea to explore the emergence of a range of categories, including, especially, multiple personality disorder, about which he continues: "Multiple personality provided a new way to be an unhappy person... a culturally sanctioned way of expressing distress" (1995a, 236). Thus understood, categories are constructed when people come to have descriptions expressing concepts of the categories that shape their actions. Provided that we accept that the instantiation of a category requires that people act under a description whose terms express the concept of the category, Hacking's action analysis can explain the emergence of categories of person alongside the emergence of concepts for them. And it can also explain the causal significance of constructed categories since these intentions and actions lead category members to be causally differentiated from nonmembers.

But despite its virtues, this action analysis is seriously incomplete as an account of category construction.[4] Consideration of its shortcomings suggests a need for a

---

[4] While Hacking (1986) sometimes speaks as though the action analysis were sufficient, elsewhere he draws attention to other elements. For instance, Hacking (1995a) draws attention to the need for diverse elements, writing that "in order for multiple personality to take off, it needed a larger cultural framework within which it could be explained and located" (41). And despite his action analysis, he remarks that "intentional action falls short of the mark... I do not believe that multiple personalities intentionally choose their disorder" (1995b, 368, cf. 1999, 115 on autism). I do not know how to reconcile Hacking's reliance on the action analysis in understanding "making up people" with these claims, but he is aware of the shortcomings of any analysis that relies exclusively on intentional action. Later, Hacking (2007) offers an account of category construction that involves both knowledge and institutional elements, gesturing at something much more like the social role account I have defended (Mallon 2003) and develop further here.

better understanding of other requirements for category construction and of other mechanisms that take part in category construction.

2.3 Beyond the action analysis

While the category constructionist understands the need for certain concepts only as a necessary condition, we have already suggested how it may also be understood as part of a sufficient one. The idea is simply that the emergence of a concept explains the creation of the category that concept is about, given a background set of other conditions that together with the concept are sufficient. What it is tempting to do, in light of this understanding, is to suppose that we know what these "background conditions" are, that they are, in effect, the normal conditions of human life or of modern life. We then can understand provocative statements about the creation of, say, race or homosexuality in the nineteenth century as the addition of the concept to a field in which these background conditions are in place, and category construction ensues. As an interpretation of constructionists, this illuminates the frequent co-pairing of claims of concept invention with category invention.

One reason to doubt the general truth of such an analysis is that: its success hinges upon claims of descriptive or conceptual novelty where this is understood as changes in the beliefs, theory, or conception that constitute the meaning of a term or concept. In Chapter 1, we saw at least that it is unclear that modern European-Americans really invented the concept of race because the most plausible substantive candidate for conceptual novelty—essentialist understandings of racial categories—is not really novel. Essentialist thinking about racial groups is not unique to modern European-American theorizing. It remains possible that *some other* claim of conceptual novelty could be sustained—if not for race then perhaps for other categories of interest. Indeed, given the possibility of opulently describing the phenomena of interest, we know that such novelty can be secured if we are willing to stipulate an opulent description to define the entity of interest.

But if we instead aim to find conceptual innovations whose introductions *explain* the emergence of important category-typical features, their novelty may be much harder to sustain. Even apparently peculiar ideas like that of multiple personalities in one body, for example, often have precursors, for instance in medical case studies and in disparate phenomena like spirit possession (Ellenberger 1970). In any case, it is far from clear that we need to make a difficult and perhaps tendentious case regarding conceptual innovation to make good on category constructionist claims. In Chapter 1, I suggested the alternate possibility that cultural innovation brought not a change characterized by the existence of some novel descriptive content, but some new distribution or social arrangement within which the content could be understood and used.

If this is the case, this new distribution or social arrangement, and not the emergence of new conceptual content, could be the key to understanding the phenomena of category construction. Call these other conditions the *social conditions of construction*. In Section 3 I offer my own account of the social conditions of construction, one that focuses upon the creation of common knowledge of categories.

Before we consider that, however, note that there is another substantial reason to doubt the adequacy of an account of category construction that focuses upon intentional action: much category differentiation is not a direct manifestation of one's choice of actions from among the descriptions and concepts that provide one with "ways to be."

Hacking's action analysis focuses especially upon *identification*, a process of seeing oneself or one's actions as falling under the category label or concept, shaping one's actions appropriately. But if what we are trying to explain is the differentiation of many different sorts of constructed categories (as I am here), then the action analysis understood in terms of identification prejudges, and judges wrongly, the source of many category differences that, plausibly, stem not only from the intentional identifications or actions of those who instantiate the categories. When we consider racial categories, sexual categories, categories of sexual identity, and so on, it is enormously plausible that many of the distinctive properties of these categories emerge as causal effects of representations that involve other causal pathways, including the actions of *others* in a community towards category members.

In addition, as a growing body of empirical evidence suggests *nonintentional* effects of representations upon the behavior of category members, and of others toward category members, play an important role as well. These sometimes unconscious, automatic processes amount to a nonobvious but potentially important layer of differential treatment that is not well captured as intentional action in the first or third person.

Finally, plausibly a great deal of category difference is not the direct result of representations modifying behavior, but rather the product of representations controlling modifications to the environment that, in turn, modify category members. Practices of racial or sex segregation, for instance, transform our physical environments in ways that feed back into the causal processes that shape members of different racial and sexual categories.

These environmental modifications, too, become penumbral under the action analysis model.

In Chapter 3, I consider such third party, nonintentional, and environmentally mediated effects of representations further. But for now, I focus upon further specifying the central mechanism of category construction: the social role.[5]

3 Social Roles

In order to frame my general discussion of the broader role that representations might play in creating causally significant categories beyond their role in structuring actions, I speak of social roles. As I use the term, a social role exists if and only if the following are met:

> SR1. *Representation*: There is a term, label, or mental representation that picks out a category of persons C, and that representation is associated with—and figures in the expression of—a set of beliefs and evaluations—or a conception—of the persons so picked out.
>
> SR2. *Social Conditions*: Many or all of the beliefs and evaluations in the conception of the role are common knowledge in the community.

Thus understood, social roles are structured by representations of categories, and the explanation for the necessity of those representations in producing those categories will involve a range of distinct causal pathways, some of which involve structuring intentional actions, some nonintentional behavior, and some environmental modification. I understand these representations as constructing social roles when the additional condition of common knowledge of the representational contents obtains in the community. And I note again that the social roles of interest to me here are *covert*: the existence, or persistence, or specific properties of the social role category are believed to be the product of natural facts, rather than human decision, culture, or social practices.

3.1 Representations

What sorts of representations are relevant to the construction of categories? While a wide range of contents can, in principle, be associated with a category, here I consider sorts of content that systematically figure in the representations

---

[5] I discuss social roles in Mallon 2003, following Griffiths 1997. Appiah 1996 pursues a similar approach, developing it further in Appiah 2005.

that structure social roles—representations of human kinds. These include information about:

> The *names* or *labels* for the category, and perhaps of contrast categories.
> *Conditions of ascription*: "If *a* has properties $P_1, P_2, P_3 \ldots P_n$, then *a* is a C."
> *Essential* features: "Necessarily, Cs are Ps."
> *Category-typical* features of Cs: "Cs are typically Qs."
> *Evaluations*: "Cs are good."
> *Norms*: "When interacting with a C, you should N." "A true C is good at P."

This list of types of content is neither exhaustive nor necessary, and these sorts of information are overlapping and not mutually exclusive. But however we specify the content of representations of categories, we need to go further than individual representations of content to understand how exactly representations result in the construction of categories, one aspect of which is the social conditions of construction.

One plausible thought regarding these social conditions that is consonant with much constructionist theorizing is that these representations produce categories by being distributed throughout a population and influencing individual behaviors and dispositions to behave. This idea is a start, but it could benefit from more precision. What amounts to the right way of sharing representations so as to result in a constructed category?

3.2 Specifying the social conditions of construction as common knowledge

I suggest that we can illuminate the social conditions of construction with the condition of *common knowledge*: information that everyone knows, and everyone knows everyone knows, and so on. "Common knowledge" is an influential idea growing out of game theory and first labeled and formulated by David Lewis (1969), and it has since received a range of sometimes nonequivalent formulations. For my purposes, I will talk of common knowledge in this sense:

> A proposition *p* is *common knowledge* in a population S if and only if
> (1) For all individuals *i*, where *i* is a member of S, *i* believes *p*.
> (2) For all individuals *i*, where *i* is a member of S, *i* believes (1).
> (3) For all individuals *i*, where *i* is a member of S, *i* believes (2).
> And so on...[6]

---

[6] "Common knowledge" is a technical term, and as we use it here, it need not involve knowledge in the philosopher's sense at all since the *kernel* of common knowledge (the believed content *p*) need not be true. The possibility of a false kernel of common knowledge is important for us since we are

Obvious objects of common knowledge include facts that are "in plain sight," truisms, and platitudes, and, more to our purposes, beliefs, evaluations, theories, and other information about human categories may also be objects of common knowledge.

Because it entails knowing what others know, common knowledge contrasts with the weaker state of *shared* belief that exists in a population when each of its members has the same belief.[7] I characterize social roles as involving common knowledge instead of the weaker condition of shared belief, because of the role that common knowledge plays in facilitating coordination and communication among different people. On this rendering, socially constructing a category involves more than just widespread ideas about the natural features of a category shaping the disparate behaviors of the population in which the idea has spread. It also involves recognition that the idea has spread which results in various sorts of coordinated, cooperative, or strategic behavior vis-à-vis the category.

That common knowledge can play such a role in coordination and cooperation is widely recognized. Consider a simple illustration of a familiar coordination problem:

Four-Way Stop: Two cars, Camaro and Skylark, simultaneously come to adjacent stop signs at a four-way perpendicular intersection. In the U.S., the rule is that the first driver to the intersection can go first, but if the cars arrive simultaneously, the driver on the right, in this case Skylark, is to go first. Suppose Camaro and Skylark know all this (they have shared knowledge of all these facts).

Camaro and Skylark still face a coordination problem. Neither wants to proceed if the other proceeds, since that could result in an accident. But at least two things about the situation make the next step unclear. Skylark knows that she has the right of way, but Skylark does not know if Camaro knows this. Camaro might fail to know this because the "yield to traffic from the right" rule is itself obscure. Or Camaro might fail to know this because it may be unclear whether the two cars really arrived simultaneously. Perhaps Camaro stopped once further back at the line and then eased forward for better visibility, stopping again simultaneously with Skylark.

---

interested in cases where widely shared beliefs (say beliefs in biobehavioral kinds or essences) may not be true.

[7] Vanderschraaf and Sillari (2009) call this *mutual knowledge*. Because this term is also used in the literature to refer to something more like *common knowledge* (Harman 1977; cf. Schiffer 1972 on "mutual knowledge"), I avoid it here. Appiah requires that social role identities be "mutually known" (2005, 67), but it is not clear whether he intends to require only *shared* belief or common knowledge.

While Skylark and Camaro have shared knowledge, they lack common knowledge, which would allow them to appropriately coordinate.

Common knowledge's primary role within game theory and philosophy has been in the rational reconstruction of coordinated and cooperative behavior by actors (e.g. Gilbert 1989). Four-way stop, above, is one example of this, and Lewis's (1969) articulation of the idea of common knowledge in his account of convention is another. This feature is important since common knowledge of a category amounts to a shared understanding that can be acted upon in certain, systematic ways by parties, even parties with diverse interests. Common knowledge of human categories enables participants in a population to share and coordinate their behavior vis-à-vis category members with the confidence that their behavior will seem rational and explicable to others.

One way of thinking about the added effect of common knowledge is in terms of *network effects*. Network effects accrue to a thing when each additional user of a thing increases its value to all users. One classic example of network effects occurs with technological devices like telephones and fax machines. These are valuable devices, but their value increases from nearly zero (when there is only one user) to very great (when there are many). The more people who have them, the more valuable they are (since the expansion of the network of users expands the power of each device). Markets (like Ebay) or social networks (like Facebook) exhibit network effects for similar reasons: the more users, the more utility these sites have for each user.

Crucially, a belief that *p* can be valuable because it tells you the way the world is. That might be useful even if no one else has the information (or, in competitive circumstances, it can be especially useful because no one else has the information). But beliefs that are common knowledge are a bit like having a telephone with a phone book that tells you who else has one. Common knowledge creates opportunities to coordinate and cooperate with others in novel and unforeseen ways. That common knowledge can facilitate more sophisticated social interchange suggests the proposal that common knowledge amounts to the right way (or one right way) of sharing representations that can result in category construction.

This proposal that the social conditions of construction are satisfied by common knowledge is not without problems. One problem is that common knowledge and related notions are often specified, as I have here, with a "hierarchical" or "iterated" structure of nested knowledge (see, e.g., Lewis 1969, 60ff; Schiffer 1972, 30; Harman 1977, 422).[8] While the iterative expression might succeed as a

---

[8] To see why these infinite, nested knowledge claims seem at all desirable, consider the role that common knowledge plays in coordinating actions. Suppose Skylark waves Camaro on. Skylark

rational reconstruction of the knowledge required for cooperation and coordination, it is unclear how it could be an explanation of real cooperation and coordination. How exactly could such an infinite hierarchy be employed by real minds in real situations? How could it be a solution to an actual problem of coordination?

Lewis (1978) seems unbothered by the potential regress the hierarchical account generates, relying instead on a distinction between implicit reasons to believe and actual beliefs (and suggesting only the former are required).[9] But there are also ways of representing common knowledge that dispense with the hierarchy.[10] A different strategy has been to attempt to sidestep the iterative expression of common knowledge with some conception of "collective belief" or "we-belief." Here, the idea is to replace the "I believe that you believe that I believe..." structure of common knowledge with a cognitively simpler "we believe that p."[11,12]

It is also important to acknowledge the idealization that common knowledge embodies. While there may be real cases of common knowledge for small groups, many real-world categories, including, near enough, all of the social categories that are of interest in the present work, contain many messy deviations from idealized cases of common knowledge. For example, the contents of common knowledge about a category can and do differ in overlapping ways within and among different social groups. For instance, what is common knowledge about categories of race, gender, or ethnicity, may be quite different among members of a category than among nonmembers. This suggests that the simplest case, where we imagine a single population participating in common knowledge, is too

---

knows that Camaro saw the wave. And Camaro knows Skylark saw the wave too (after all, Skylark made it). But suppose that Camaro doesn't know that Skylark knows that Camaro saw the wave. Then Camaro won't go. After all, if Skylark thinks Camaro did not see the wave, then Skylark might think that Camaro thinks Skylark has the right of way. In this situation, Camaro could imagine that Skylark, believing that Camaro believes that Skylark has the right of way, will go first. Because doubts anywhere down the infinite chain of belief seem to undermine the knowledge needed for cooperation, and because cooperation does happen, the infinite hierarchy seems necessary if ordinary cooperation is to be rationalizable.

[9] See also Peter Klein's spirited articulation and defense of "infinitism" in epistemology (1999).

[10] For example, Gilbert Harman has suggested the relevant idea (which he calls "mutual knowledge") "might be explained as knowledge of a self-referential fact: A group of people have mutual knowledge of p if each knows p AND WE KNOW THIS, where the THIS refers to the whole fact known" (1977, 422).

[11] Searle (1990, 1995, 2010) makes a similar move in the context of understanding collective intentions.

[12] To add even more complexity, exactly how we represent common knowledge may differ for different situations, for example situations in which we have common knowledge of an object of joint perception or attention, as opposed to common knowledge of a shared cultural platitude. See, e.g., Gilbert (1989); Clark (1996); Tomasello (2008) for discussion of such differences.

simple, and instead we must imagine overlapping populations of people constructing overlapping social roles that may sometimes subject putative category occupants to inconsistent or even practically contradictory expectations.

Here I put debates about the exact representation of common knowledge to the side, and continue to focus on the simplest case of common knowledge while allowing that it is an idealization. My aim is to offer a core mechanism for a "how possibly" explanation for category construction. I assume that if we can do so for an idealized case, then we can add whatever complex extensions, qualifications, and emendations to the idealized account that are needed to address actual situations, as appropriate.[13]

3.3 Covert social roles are not conventional or institutional kinds

We have emphasized the role that mistaken belief in naturalness plays in constructionist theorizing, noting the revelatory aim of constructionist accounts. While it is common to regard socially constructed social roles as analogous to conventional or institutional social roles like *being married* or *being a member of parliament*, they are fundamentally different because they are made or sustained in part by *mistake* rather than by strategic cooperation.

To highlight this contrast, consider Lewis's (1969) seminal account of conventions. Roughly, on Lewis's account, a convention is sustained when it is true that, and it is common knowledge that, everyone participates in a behavioral regularity on condition that everyone else does (1969, 76). Crucial to Lewis's account of convention is that a conventional conformity is a *Nash equilibrium*, a regularity that no one can unilaterally deviate from and better satisfy her own preferences. The stability of the regularity depends upon this fact (and on common knowledge of it). In contrast, much behavior with regard to a putatively natural category is not strategic: it is not conditional on the way others will act; it is conditional on the way the world is taken to be.[14] This stabilizing mechanism distinguishes covertly constructed categories from overtly constructed ones.[15]

---

[13] This account uses common knowledge as an idealization in yet another way, for only some of the avenues of construction that I discuss in Chapter 3 require common knowledge. For others, a weaker condition might do. In these cases, speaking in terms of common knowledge is a way of specifying a sufficient condition.

[14] For similar reasons, covert social roles are not institutional facts of the sort John Searle (1995, 2010) analyzes. On Searle's account, social reality is produced by a community collectively "imposing function" upon an individual or type. In contrast, social roles in the present sense are produced simply by common knowledge of putatively natural facts about the world. I discuss this example further in Chapter 8, Section 2.1.2.

[15] Note that this is a *synchronic* claim about constructed categories, not a genealogical claim. That is, it could well be that a category began as a social convention or institution, but then came to be

3.4 Common knowledge and public broadcasts

Thinking about the social conditions of category construction as a condition on *how* categories are represented illuminates claims of category invention. Begin with the idea that common knowledge is produced by events that are *public*, or *open*, or that *broadcast* information to a group. Consider that right of way at a four-way intersection is normally established by judgments of readily observable facts and publicly disseminated rules: for example, by temporal order of arrival and the "right goes first" rule when arrival is simultaneous. When those don't work, as in the case we considered in Section 3.2, further public signals are required. Perhaps Camaro signals Skylark by waving, or Skylark attempts, quite visibly and slowly, to ease ahead. All of these are public facts that are broadcast to a population.

Some broadcasts are *public* in the specific sense that when a public broadcast that *p* occurs, in addition to acquiring content *p* (and perhaps coming to believe it), a receiver *r* of the broadcast also acquires the information that other receivers have acquired *p*, and also that other receivers understand that *r* has acquired *p*, and that *r* has acquired the information that other receivers know this. And so on. That is, a broadcast is public in this sense if it causes the broadcast content to become common knowledge among the population of receivers. The rules for right of way are supposed to exploit public broadcast of the relevant facts (who arrived at the intersection first, the rules of priority, etc.) in order to ensure successful coordination.[16] In my example above, where this broadcast was in doubt, new public broadcasts (waving, or conspicuously easing one's car into traffic) are possible solutions.[17]

Consider again the case of the invention of racial categories. In Chapter 1, I conjectured that specific lineage essentialist conceptions of particular races

---

regarded as natural and inevitable. The point is that once a category is regarded as natural and inevitable, it is a different sort of kind than a conventional social role.

[16] Whether a broadcast is public for a population of receivers depends on the background common knowledge that the receivers have. It's because we mutually expect one another to perceive and reason about events in the same way as we do that certain kinds of events count as public broadcasts. In this way, the acquisition of common knowledge already depends on a background of common knowledge. This suggests that broadcasts cannot be the only source of common knowledge. Indeed, it may suggest that some rudimentary common knowledge is part of the species-typical cognitive endowment of the hyper-social human animal.

[17] Lewis's own account of common knowledge foregoes a hierarchical characterization of the sort I sketched above in favor of a characterization that suggests a certain conception of the source of common knowledge—some state of affairs *A* that is apparent to members of a population and that indicates something to them. Similarly, Margaret Gilbert's (1989) account of common knowledge involves the idea of some fact being *open** to a population, where the conditions on "open*" are attempts to specify the conditions of what I have called a public broadcast.

came to be widely promulgated by the emergence of an international scientific community. One way of understanding this process is that scientific institutions produced a public broadcast that produced common knowledge of these beliefs among the broad and growing community of consumers of science. The result would then be common knowledge of the beliefs producing what I have called a social role.

Similarly, remember Foucault's claim that homosexuality appeared "when it was transposed from the practice of sodomy into a kind of interior androgeny, a hermaphrodism of the soul" (1978, 43). One can see again here the suggestion that a nineteenth-century shift occurred in which homosexuality came to be explained in terms of inner essences that demarcated newly discovered sexual categories, and explained their typical properties. But Foucault also emphasizes the role of scientific discourse in bringing such a transformation about, as when he writes, "the psychological, psychiatric, medical category of homosexuality was constituted from the moment it was characterized" (1978, 43). The public broadcast of beliefs transforms the social world by producing common knowledge about a putative category, thereby producing a social role.

To be clear, the role of scientific institutions in producing credible public broadcasts is not the only source of common knowledge about social roles. A somewhat different sort of public broadcast is suggested in Hacking's work on multiple personality disorder. Multiple personality has been shaped by culturally local theories of the category both within and across cultures, as Hacking and others have argued (e.g. Spanos and Burgess 1994; cf. Ellenberger 1970). One sort of evidence for this comes from the fact that as beliefs about the category became widespread, so did patients with the diagnosis. And as the specific conception of the diagnosis changed, so did the symptoms of the patients. For instance, as the 1980s wore on, theorists began to posit more and more personality fragments per patient, a hypothesis that was subsequently born out in the manifestation of increasing numbers of personality fragments for each patient (Hacking 1995a, 77).

Many have suggested that multiple personality disorder may be iatrogenic, or induced by therapists. (And, indeed, there is evidence that one prominent case—Dr. Cornelia Wilbur's famous patient Sybil—may have been a case of medical fraud (Rieber 1999).) However, Hacking argues for a more complex account of the rising tide of diagnoses: that the category came to be promulgated through the mass media, so that patients began to appear in doctors' offices with their symptoms already in tow, seeking some social recognition or benefit. Mass media, talk shows, and popular culture are powerful mechanisms for producing common knowledge that are somewhat distinct from the institutions of science.

But a common knowledge understanding of the category could allow patients to exhibit behavioral symptoms, knowing that others will recognize their symptoms as indications of category membership and coordinate their responses accordingly.

The connection between public broadcasts and common knowledge also illuminates another important feature of social categories, the fact that one's membership in some categories may be publicly broadcast or *conspicuous*. Some categories are *volitionally* conspicuous in this way. Some cultural groups broadcast membership by wearing clothes or other indicators that mark one as a member of the group. And such conspicuousness can be directed at everyone, or only at other group members who may share a special store of common knowledge about the indicators. But many other categories are *nonvolitionally* conspicuous. These involve, especially, categories indicated by the body like race and sex (as well as very different ones like dwarfism or Down Syndrome). Still other categories, like language group or ethnicity or class, can fall in between.

Theorists of race have long considered the importance of skin color in understanding the situation regarding American racial groups.[18] Where conventionally assigned race is conspicuous, it can be difficult for an individual to elude social role expectations and restrictions. At the same time, nonvolitionally conspicuous membership can be a difficult-to-fake marker of in-group membership. In this vein, some groups use bodily modifications to produce conspicuous, hard-to-change indicators of membership. Modifications to one's body, hair, skin, or face, for example, can have significance stemming in part from the fact that they make one's commitment to a certain identity or way of life conspicuous and more difficult to elude.

4 Social Roles and Causal Significance

I have been discussing common knowledge by way of specifying the social conditions of category construction. But for all I have said so far, the notion of a social role still looks too weak to understand how constructed categories might be causally significant. Social roles, understood as existing wherever there is common knowledge of a category, are ubiquitous, and often insignificant. Consider the category *person whose last name begins with the letter Q*. Or the category of the *prematurely bald*. Or the category *hybrid car owner*. Each of these categories carries ideas about the sort of people who occupy it, and supports

---

[18] E.g. Alcoff 2006; Hardimon 2003; Taylor 2013; Omi and Winant 2014. See Ignatiev 1995 and Brodkin 1998 for accounts of how Irish and Jewish persons, respectively, came to be regarded as white.

at least a few generalizations about them. (E.g. "Persons whose last name begins with the letter Q will line up alphabetically before the Ws but after the Gs.") But these categories, social roles though they may be, fail to be causally significant.

To be sure: exactly how causally potent a kind must be in order to count as an important kind depends crucially on the exact nature of our explanatory and epistemic projects. For *some* purposes, even these thin social roles may be causally efficacious enough to make them relevant kinds. But it remains that the category constructionist needs to understand how social roles of this minimal variety, roles that we normally have no interest in picking out as kinds, could become so causally significant as to make reference to them crucial to understanding our social world. What is needed is some understanding of how common knowledge representations of a category might construct the category in some more substantial sense, how they might become what I call *entrenched* social roles.

3

Social Roles that Matter

What makes some constructed social roles matter? Explanations of the construction of a particular category can mean offering a very specific account of the causal, historical, social, and psychological processes or mechanisms by which terms and concepts of that category emerged and came to shape the causal structure of the world. The local, specific nature of much constructionist research bears this out, revealing extraordinarily complex, historically and culturally contingent circumstances in which representations come to have the content that they do, and categories come to be the way they are. Given this, one might consider the individual and contingent historical mechanisms appealed to by such specific explanations and conclude: "that is all there is; there is no more." That is, one might hold that there is nothing very general to be said by way of specifying the content or mechanisms of category construction.

In contrast to this explanatory particularism, I characterize category construction in a general way in order to illuminate a whole range of category constructionist claims. In particular, I explore three different pathways by which common knowledge representations of a category can causally differentiate the apparent members of the category, becoming causally significant, entrenched social roles. These pathways include structuring intentional action by and toward category members, shaping the manifestation of automatic cognitive processes by and toward category members, and guiding behaviors that shape and reshape the world we live in. My aim is to articulate an account of the construction of causally significant human kind categories that is both general and possible. I do not insist that such a general level of explanation captures all that is causally important in episodes of category construction, but simply that there are some common mechanisms in play and that we can benefit from attempting to characterize them and their varieties.[1]

---

[1] In this way, the discussion is in the philosophical tradition of work by Linda Alcoff, Kwame Anthony Appiah, Michel Foucault, Paul Griffiths, Ian Hacking, and Sally Haslanger, more than in the humanistic social scientific traditions that guide much constructionist research. Cf. Griffiths

These mechanisms are such that, if realized, they could make category constructionist explanations true, and that their realization is (in some cases) compatible with what we know about the natural world in general and human nature in particular.

The result is something like a "how possibly" model for explaining the causal power of constructed kinds (Brandon 1990; Dray 1957), but with an important qualification. I am not offering a "how possibly" explanation of any specific category, but a range of parts out of which such explanations can be constructed for specific categories. These parts will not all be implicated in every case, nor do I claim that the co-realization of all of them in an individual case is possible. Rather, I take it that, given these parts, it is easy enough to see that a range of how possibly (and even "how actually") explanations can be built that make specific human categories that causally matter.

In Sections 1 and 2, I consider a number of sorts of ways that the existence of social roles of the sort introduced in Chapter 2 can influence behaviors by and toward category members. Specifically, in Section 1, I return to consider the production of intentional behaviors in response to category representations. Then, in Section 2, I consider a range of evidence for nonintentional, sub-rational, psychologically *automatic* influences on behavior of common knowledge representations—influences that are not mediated by intentional, conscious processes like reasoning. In Section 3, I briefly consider several sorts of *environmental construction*: alterations to the environment that alter the context for oneself and others. In Sections 4 and 5, I return to articulating an account of the causal significance of such entrenched social roles, arguing that behavioral consequences of social roles and their lasting effects on the environment can produce what Richard Boyd has called "homeostatic property clusters"—clusters of nonaccidentally co-occurring properties that sustain explanatory and practical projects including prediction, explanation, and intervention.

1 Behavioral Influences: Intentional Action

In Chapter 2, I discussed Hacking's action analysis that understands category constructions as produced by the possibility for intentional actions framed by meaningful descriptions. Hacking's action analysis follows long humanistic traditions of theorizing about the social world that explain human behavior as an intentional and rational response to ways that we construe the world. On this

(1997, chapter 6); Murphy (2006); Kuorikoski and Pöyhönen (2012); and Drabek (2014) for other attempts to characterize some empirically supported, general mechanisms of construction.

picture, representations influence behavior because they represent the world (including oneself) in certain ways, as having some properties rather than others, thereby modifying the behavior of community members in rational ways.

Hacking writes of "making up people" and the "looping effects of human kinds"—the way in which the production of representations of kinds by social scientists sets in motion a range of activities that result in the creation of categories and the transformation of properties of category members, leading, in turn, to further changes to the representations (Hacking 1986, 1995a, 1995b, 1998). In the case of multiple personality disorder, Hacking argues that spreading beliefs about a category of person creates opportunities for distressed persons to perform category symptoms allowing them to achieve or acquire certain social benefits, and this spread also created opportunities for therapists and clinicians to identify, care for, and extend knowledge of new instances of a category.

For Hacking, this process involves both a willingness of those who use some category labels to recognize and favorably respond to behaviors appropriate to the role, and a willingness of candidate category members to perform behaviors appropriate to the role. On this view, category members intentionally perform the behaviors as represented by the theory of the category that is common knowledge in the community, and therapists and others recognize the performance as such, again because of common knowledge of the category—of its typical manifestations and of appropriate responses to them. Common knowledge is crucial to this coordination since it enables would-be multiples to know how their actions will be interpreted by unknown therapists.

K. Anthony Appiah's work on social identities (identities like "black and white, gay and straight, man and woman") similarly emphasizes an intentional, rational role for category members:

Once labels are applied to people, ideas about people who fit the label come to have social and psychological effects. In particular these ideas shape the ways people conceive of themselves and their projects... the label plays a role in shaping the way the agent makes decisions about how to conduct a life. (2005, 66)

Their emphasis on the intentional, rational character of constructed behavior suggests that both Hacking and Appiah conceive of some category occupants as more or less rational agents who, given their beliefs about their environments and themselves, choose actions in accord with their interests, given their circumstances.

Other social theorists similarly seem to suggest something like a rational actor. Judith Butler's talk of gender "performativity," for instance, or Jean-Paul Sartre's (1956) radical emphasis of agency in the production of apparently natural behaviors, or James Averill's performative explanations of emotions (1980a, 1980b, 1994) exhibit a similar emphasis on explaining behavior as a rational, intentional response to a social situation that is shaped by common knowledge representations.

The notion of "rational" here is normatively weak; it implies nothing about the prior justification of the beliefs and desires that figure in producing the action, nor anything about the justice or fairness of the states of affairs they represent, for example. Allowing that social role behavior is weakly rational does not suggest that it is uncoerced or otherwise good. When the robber brandishes a weapon and asks for your smartphone, it can be rational to hand it over. Less dramatically, one's social role may be what Appiah elsewhere calls "tightly scripted"— giving a category member few options to shape her own life—but one's actions can still be intentional, rational responses to it (1996, 99). Even when we allow that behavior by representation users is weakly rational, there are multiple different (though not mutually exclusive) ways that social role representations may lead to intentional, role-differentiated behavior.

As we saw in Chapter 2, Hacking emphasizes the role that a new representation of a category can produce in promulgating the category as a way that a person could be, suggesting to members of the community various possible actions and identities. And while Hacking tends to focus upon the ways that new labels can provide salient possibilities *for* category members, common knowledge of categories may also include conceptions of salient actions *towards* category members.[2] To this salient scripting role for representations, we can add two other dimensions along which representations rationally influence intentional actions.

Nonstrategic actions occur in virtue of belief in the content associated with the representation of the category offering a reason for action, and *strategic* actions exist in virtue of beliefs about what others will do, given common knowledge of the category. "Strategic" is used here to refer to choices that are conditional upon how others will choose. Recognizing that a person falls into a category C about which one has a set S of common knowledge beliefs may rationally influence one's behavior because,

Salient Possibility: S suggests a way of acting *a* as a possible or salient choice.

Or because

Nonstrategic: Salient Possibility is true. S represents the intrinsic properties of members of C as *q*. That Cs are *q* figures as part of a reason for acting in way *a*.

---

[2] Similarly, Appiah 2005, 68f; Murphy 2006, 265; and Sveinsdóttir 2013 all register a central role for third-party classification.

Or because,

> *Strategic*: Nonstrategic is true and common knowledge. That Cs are q is common knowledge figures as part of a reason for acting in way a.

Each of these causal pathways can operate upon both the labelers and the labeled: on those who use the descriptions to think about others, and those who use them to think about themselves.

1.1 Salient possibility

In Chapter 2, we considered Hacking's action analysis on which the availability of new concepts makes actions under intentions constituted by those concepts possible and salient. To add another example, Hacking suggests that, "one thing that some pornography does is to disseminate new modes of action, new descriptions, verbal or visual" (1995a, 239). Hacking suggests this is harmful since "most men, including many cruel and abusive men, are remarkably innocent, being simply unacquainted with the range of possible demeaning actions" (239). Hacking suggests that some pornography extends to cruel and abusive men new ideas for abusing. Defenders of pornography might instead emphasize the prospect that new ideas could offer more possibilities for satisfying actions. The shared idea—that media can produce new salient possibilities for action and that this has actual effects on behavior—seems confirmed by emerging recognition of the role that now ubiquitous pornography is playing in changing expectations about what sex should be like.[3]

1.2 Nonstrategic reasons

Common knowledge representations of putatively natural kinds also provide nonstrategic reasons for actions via their representation of the properties of the kinds. Our beliefs of what we take to be the structure of the world give us nonstrategic reasons to act in particular ways in order to advance our interests. These beliefs include beliefs about the various kinds we take humans to instantiate. The sexist employer who uses sex as a proxy for filling jobs of a type on the grounds that she believes most or all members of a sex are, in virtue of their biology, unsuited for work of that type is engaged in nonstrategic, weakly rational behavior.

---

[3] Marston and Lewis 2014.

Such behavior has a parallel for self-directed representations, for the theories we hold about what kinds we belong to offer us guidance as to what we may or may not be successful at, what the background conditions of our projects and goals might be. Carol Dweck has argued that how individuals think about their capacities—as either fixed, or capable of growth—influences their performance (Dweck 1999). For instance, students with a "fixed" mindset interpret difficulty in performing a task as evidence for their lack of innate talent, performing worse and withdrawing sooner from such tasks. In contrast, students with a "growth" mindset interpret difficulty as an opportunity to grow their capacities and ultimately achieve greater success. Importantly, in withdrawing effort, attention, and identification from such tasks, Dweck's "fixed" mindset subjects may be mistaken, but they are not irrational. If a task simply promises to provide more evidence about what one believes is one's unchanging lack of ability, withdrawing from it seems entirely rational.

This suggests that to the extent that membership in the putatively natural categories of interest to covert constructionists is seen to reflect the assumption that category members' abilities in a domain are fixed, self-categorization can rationally lead one to withdraw effort, attention, and identification from the task. And the folk essentialist construals of categories we discussed in Chapter 1 provide one set of cognitive dispositions that might facilitate such assumptions.

Alternatively, representations of a category as natural might modify our behaviors by modifying our moral judgments and retributive attitudes towards category members for category-typical behaviors. Consider, for instance, contemporary social moral discussions about whether homosexuality is natural or an "orientation," or in contrast, is a choice or preference. What is being negotiated with these claims is, in part, whether homosexual behavior is an appropriate target of praise and blame, and the fulcrum of this debate is a presumed connection between being natural and being involuntary. I explore this connection further in Chapter 4.

1.3 Strategic reasons

In addition to these nonstrategic reasons, social roles also offer *strategic* reasons for actions, reasons that obtain because of common knowledge in a community. Common knowledge about a category allows us to reason about how others will act as we choose our own actions. Remember now our sexist employer, but now imagine the employer has no beliefs that, say, men are inherently inferior at caregiving professions. She may nonetheless know that others in her community are sexist and so know that hiring, say, a man into a caregiving position will result in community members taking their business elsewhere, and perhaps even insisting that others do as well.[4] She may thus rationally refuse to hire a man even if she does not believe the widely held sexist representation.

Again, such strategic behavior can also modify behavior by category members. Hacking's work on multiple personality disorder, for instance, implies that one reason some people act as a multiple is because of the affordances to achieve recognition and support from others that the common knowledge representation of multiple personality disorder offers them.

Strategic explanation is also at the core of Signithia Fordham and John Uzo Ogbu's (Fordham and Ogbu 1986) controversial theory of "acting white." The theory of acting white attempts to explain black–white gaps in academic performance by appeal to the idea that African Americans have an "oppositional culture" that represents academic performance as "acting white"—a culture in which academic success signifies a betrayal of, or attempted defection from, black identity. The theory has been influential. A study in North Carolina (Tyson et al. 2005) showed that many educators believe it (594). It has even received repeated mention from President Obama.[5] Quantitative evidence for the theory has been hard to come by (for instance, Tyson et al. (2005) found little evidence for racialized oppositional culture in North Carolina schools), and a range of evidence points in the opposite direction (Harris 2011; Tyson 2011). A study by Roland Fryer and Paul Torelli (2010) found evidence consistent with the theory for some students in some U.S. high schools, but their data apply, at best, to a small percentage of all black students in the U.S. (Harris 2011, 23-4, cf. chapter 5). It seems as though while such oppositional culture may exist in some people and places and therefore has some explanatory role to play, it does not explain a large part of black/white differences in educational outcomes in the United States.

But the "acting white" hypothesis does illustrate the possibility of appealing to rational, strategic behavior in the explanation of widespread differences attending everyday social categories. The theory has it that students respond to common knowledge representations connecting racial identity and academic performance in their communities. Where this occurs, students are engaged in strategic, rational behavior with regard to racial identities, for it can be rational for a person to discount the benefits of better academic performance if it comes at social cost. And such strategic calculations concerning the interaction of social

---

[4] This would amount to her not endorsing the belief at the core of common knowledge in her community, and so not fully participating in the common knowledge.

[5] E.g. in his 2004 Democratic National Convention Speech and again in July 2014 at the Walker Jones Education campus, a preschool–eighth grade in Washington, DC.

understandings and the behavior of others can remain important whether or not one believes in the content of stereotypical representations of the role.

Weakly rational explanations of social role behavior thus may include the role that concepts play in making certain action possibilities salient, and also both nonstrategic and strategic reasoning about human categories and one's community. And this sort of constraint may occur both among those who use the labels and concepts and among those labeled as belonging to the category.

2 Behavioral Influences: Automatic Processes

As I suggested in Chapter 2, focusing on intentional action alone misses a range of ways that representing a category can construct a category, including ways in which such representing does so rather directly, by altering category-relevant behaviors. In this section, I focus on some recent evidence of such effects that focuses upon what are often called *automatic* processes. Over the last two decades, there has been growing attention in psychology to such automatic processes, processes that are activated by representations of situations and proceed quickly and without drawing on complete background knowledge or on the "executive" resources of conscious attention, reflective reasoning, or willpower. Correspondingly, the behavioral outcomes of such processes may deviate considerably from those that would emerge from controlled processes that, because they draw on more background knowledge and "executive" resources, are slower though comparatively smarter and more flexible. Automatic processes may compete with these slower processes for control of behavior.[6]

By most accounts, automatic processes are a motley collection, comprising a wide range of different mechanisms, including perceptual processing, linguistic processing, autonomic processes, and a range of other processes, including a range of domain-specific sorts of processing that seem to be functionally specialized adaptations for specific tasks. Candidate specializations might include reasoning about others' mental states (e.g. Leslie 1994), or about the biological world (e.g. Atran and Medin 2008; Barrett 2005; Keil 1989), or about potential contaminants (e.g. Kelly 2011), or about norm violations (Sripada and Stich 2006; Chudek and Henrich 2011). Some automatic processes are plausibly "domain-specific" solutions to long-term adaptive problems, but the task of coping with

---

[6] Bargh and Chartrand 1999; Gendler 2008; Greene et al. 2001; Haidt 2001; Kahneman 2011; Stanovich 2004.

human categories is surely one that cuts across many cognitive domains, activating and incorporating a wide range of mental mechanisms and behavioral dispositions in complex ways.

2.1 Mere distinction

Among the most surprising effects of representations upon behavior concern not the content of what is represented about a human category, but rather *mere distinction*. Merely representing a category using a distinct marker seems to trigger automatic cognitive propensities to treat such markers as indicators of important social categories. Consider a few examples.

Henri Tajfel and colleagues' classic social psychological work on so-called *minimal groups* sought to discover how little in common group members could have before they showed preference to in-group members over out-group members. One famous early study had experimenters divide a group of students from a boys' school into arbitrary categories, and then offer them a chance to distribute real monetary rewards to other boys from the school (but whose specific identity they did not know). Tajfel describes the experiment:

> The boys, who knew each other well, were divided into groups defined by flimsy and unimportant criteria. Their own individual interests were not affected by their choices, since they always assigned points to two other people and no one could know what any other boy's choices were. In as much as they could not know who was in their group and who was in the other group, they could have adopted either of two reasonable strategies. They could have chosen the maximum joint-profit point of the matrices, which would mean that the boys as a total group would get the most money out of the experimenters, or they could choose the point of maximum fairness. Indeed, they did tend to choose the second alternative when their choices did not involve a distinction between in-group and out-group. As soon as this differentiation was involved, however, they discriminated in favor of the in-group. (Tajfel 1970, 101)

Tajfel's startling discovery was that randomly assigning students to arbitrary, artificial groups that were unconnected with their self-interest led them to favor in-group members over out-group members, even at cost to others from their school.

More recent work tells the same story about minimal groupings. For instance, in a study of five year olds, Yarrow Dunham, Andrew Scott Baron, and Susan Carey (2011, experiment 1) found that five-year-old children randomly assigned to minimal groups (the red group and the blue group) exhibited in-group favoritism on both explicit judgments (a ranking of liking or disliking an in-group or out-group person) and implicit measures of attitudes (measured by response times in pairing positively or negatively valenced spoken words with in-group or out-group members) and resource allocation (giving money to an in-group or out-group member).

In a quite different study on undergraduates, Robert Kurzban, Leda Cosmides, and John Tooby (2001) reported mere differences of clothing color affected the organization of information in memory in ways that outran the effects of race. Kurzban and colleagues suggestively linked these clothing markers to distinct teams or "coalitions"—hypothesizing that such a link would drive automatic categorization, though later work suggests that even the clothing markers were inessential (Peitraszewski et al. 2014).

In Chapter 1, we considered several strands of psychological literature that also suggest that people impute considerable significance to minimal differences.[7] We considered at some length the literature on broad essentialist reasoning about a range of kinds, and lineage essentialist reasoning about human racial or ethnic groups. Such essentialist reasoning takes very little to activate. Recent developmental evidence from Marjorie Rhodes, Sarah-Jane Leslie, and Christina Tworek (2012) suggests the existence of a predisposition in children to reason in an essentialist manner about novel kinds merely given novel generic labels (e.g. "Look at this Zarpie! Zarpies are scared of ladybugs!"). Interestingly, they also found a corresponding willingness of adults to use generic labels to communicate information about kinds that have been characterized in an essentialist manner, suggesting that children and adults alike connect essentialist assumptions with mere use of generic terms to describe a group. In Chapter 1, I also noted Katherine Kinzler and colleagues' evidence of accent-driven playmate preferences in young children, preferences that emerge well before effects of race on those same preferences (Kinzler et al. 2009). These preferences are driven by mere accent and not the content of what is said.

Each of these cases is an example of somewhat minimal distinctions that automatically trigger processes with real effects on evaluation, cooperation, memory, and reasoning. One way of characterizing their common thread is that each represents a different sort of confirmation of the "tribal instincts hypothesis"—the hypothesis that human cognition has been evolutionarily adapted for living in extended cultural groups characterized by "tribes"—groups that coordinate, cooperate, and share social norms and institutions (Richerson and Boyd 2005, chapter 6; Kelly 2011; Greene 2013). In such a context, cognitive tendencies that allow one to discern those with whom one can and cannot expect to share interests and norms could be valuable in avoiding the wasted energy that

---

[7] For recent neuroscientific work emphasizing minimal differences, see Cikara and Van Bavel 2014.

comes from failures to coordinate and cooperate successfully (Henrich and McElreath 2003). In this light, these predispositions to jump to conclusions about people on the basis of very little information make evolutionary sense. While the truth of this evolutionary hypothesis is not essential to this proximal account of the social construction of causally significant human categories, it does allow a better understanding of the fit between evolutionary-cognitive and social-constructionist theorizing about human thought and behavior. In Chapter 7, Section 4 I return to this idea by way of making the case for the possibility of stable entrenched social roles.

2.2 Dissociations between automatic and rational processes

One striking feature of many automatic processes is that because they operate automatically, and sometimes unconsciously, they may have effects on our behavior that run counter to our explicit intentions, plans, and projects. (This is why, for instance, the study by Dunham and colleagues (2011) on minimal group effects on five year olds, mentioned in Section 2.1, measured both.) A pair of examples—one concerning users of labels, and the other those who fall under a label—demonstrate this.

A burgeoning literature in social psychology documents the existence of so-called implicit attitudes that shape behavior on indirect measures (e.g. disjunctive categorization under time pressure characteristic of the "Implicit Association Test" (IAT) or proximity of chair placement to a person of another category (as in Amodio and Devine 2006, study 3)) but that may not show up on explicit measures of attitudes such as simply asking a person.

So, for example, the Implicit Association Test measures how quickly subjects respond, and how likely they are to make errors, on categorization tasks (Greenwald et al. 1998). In one task, subjects are asked to categorize using disjunctive categories (e.g. "African American or good," "European American or bad") either images of members of a human category (e.g. pictures of black/white faces) or positively or negatively valenced words (e.g. "wonderful" or "nasty"). Such tests find that American subjects categorize more quickly when white faces are conjoined with positive (rather than negative) words, and more slowly when black faces are conjoined with positive (rather than negative) words (e.g. Nosek et al. 2002a). These results suggest the presence of preexisting associations between black faces and negative evaluations and white faces and positive evaluations, associations that facilitate faster responses on the test for pairs consonant with them. The growing literature on implicit biases has shown similar biases with regard to categories like body type (Teachman and Brownell 2001), age group (Nosek et al. 2002a), and sex (Nosek et al. 2002a, 2002b). What is striking in all this is that indirect measures seem to reveal negative attitudes towards members of a category even when explicit attitudes do not (Nosek et al. 2002a; Teachman and Brownell 2001).

While there is ongoing debate about the real-world implications of such implicit biases, a growing body of evidence suggests real-world consequences.[8] For example, implicit associations have been found between African Americans and weapons (Nosek et al. 2007), and other behavioral studies suggest a "black-weapon bias"—a bias to perceive a neutral object primed by the face of a black person as a weapon in a task requiring rapid categorization (Payne 2001)—a finding with obvious real-world implications for understanding decisions in policing. Implicit biases may also play a role in employment discrimination (Steinpreis et al. 1999; Bertrand and Mullainathan 2004), in the refereeing of sporting events (Price and Wolfers 2010), and in judgments of scientific ability or potential (Moss-Racusin et al. 2012). The dissociation between implicit attitudes (or attitudes as indirectly measured) and explicit attitudes suggests that *even representations that are not believed or desired* can influence one's behavior in ways that have practical effects.

This contrast between the behavioral outcome of processes that sometimes control one's behavior and one's all-things-considered attitudes suggest that considering the influence of category representations on behavior vis-à-vis category members exclusively in terms of intentional action would be to ignore a persuasive and growing body of evidence that documents representational influences on behavior that are different than, and sometimes at odds with, agents' intentions.

Are similar dissociations between automatic and controlled processes possible in the case of *self-directed* representations by the labeled? Some models of *stereotype threat* effects on performance suggest the answer is "yes." Stereotype threat is the threat each of us faces in a situation in which our behavior or performance might be interpreted as confirming a stereotype about a group to which we belong. Evidence suggests that subtle primes of categories subject to stereotype—including race and ethnicity (e.g. Steele and Aronson 1995; Gonzales et al. 2002), gender (Shih et al. 1999), and socioeconomic status (Croizet and Claire 1998)—can degrade performance on a range of tasks including athletic endeavors (Stone et al. 1999; Stone 2002), managing one's impressions on others (Frantz et al. 2004), and, notably, academic examinations (e.g. Steele and Aronson 1995; Shih et al. 1999; Shih et al. 2002).[9] Interestingly, these effects

---

[8] For a competing interpretation, see, e.g., Oswald et al. (2013). The present discussion rests on the idea that, as Greenwald et al. (2015) and others (e.g. Valian 1998; Mallon and Kelly 2012) have suggested, small effects may accumulate over time to produce substantial differences.

[9] A range of social psychology results have come under increasing scrutiny in recent years, fostering systematic attempts to replicate a range of studies. One of these attempts, Moon and

have the shape of what Robert Merton called a "self-fulfilling prophecy," wherein stereotypes that represent, say, blacks as bad at academic tasks, whites as bad at athletic tasks, or women as bad at mathematical tasks actually bring it about that members of these categories perform worse in the task domain.

While there is no consensus on the exact explanation of these effects, some models suggest a dissociation between the representations that drive stereotype threat effects and explicit attitudes. For instance, one "stereotype imitation" model has it that stereotype threat occurs when subtle primes activate stereotypes that represent category membership deleteriously with respect to performance in a domain. These primes, in turn, facilitate or produce behavior enacting the deleterious stereotype—an imitative phenomenon that seems exhibited in other well-known studies on priming (e.g. Bargh et al. 1996).[10] Crucially, on this imitation model of stereotype threat, it seems possible for a subject to be susceptible to stereotype threat even if they do not believe the deleterious stereotype.

A different, more prevalent model suggests that recognition of a situation as threatening imposes additional cognitive burden that in turn interferes with performance in task domains (e.g. Steele and Aronson 1995; Schmader et al. 2008). Such a recognition plausibly exploits something like common knowledge of the representation and the significance of a behavior in the situation (knowledge of what a certain kind of performance would be taken to mean by others).[11] As with the imitative model, recognizing a situation as posing a threat is consistent with a failure to believe the stereotypical association is true.

2.3 Acquired automaticity

Some theorists of automaticity interpret automatic processes in the wake of evolutionary psychology and emphasize innate, domain-specific mechanisms that operate automatically in processing solutions to stable, recurring problems of human natural history (e.g. Stanovich 2004). But it is clear that learned or practiced behaviors may also become "stacked" and automatic, facilitating

---

Roeder (2014), failed to replicate Shih et al's (1999), but another, Gibson et al. (2014), did replicate the effect.

[10] Bargh, Chen, and Burrows's influential study, and the imitative model it develops, has prompted failures to replicate key experiments by Doyen et al. (2012) and Pashler et al. (2011). Others have replicated the effect, but not supported the imitative interpretation (e.g. Cesario et al. 2006). Chartrand and Bargh (1999) offer other evidence for automatic imitation, calling it "the Chameleon Effect."

[11] Strictly speaking what is required is knowing that others will take poor behavior in the domain as confirming the deleterious stereotype. So common knowledge is a stronger condition than is required, but it is a perspicuous way to state a sufficient condition.

sophisticated behaviors without conscious awareness, and this is true even where those behaviors override innate, automatic predilections. One nice example emphasized by Paul Griffiths is that of "display rules" (1997, 156)—cultural rules that modify emotional displays. Drawing upon work by Paul Ekman, Wallace Friezen, Carol Izard, and others on the so-called "basic emotions"— the five or six emotions (usually including anger, happiness, sadness, fear, and disgust) that are paradigms of evolutionary psychological success—Griffiths points out that even these apparently "biologically determined" motor responses can be shaped by cultural reinforcement; for example, Ekman et al. (1972) showed that Japanese, but not American, students would suppress their "automatic" emotional facial expressions in the presence of an authority figure, suppression that took place so quickly and automatically so as to be nearly undetectable (with Ekman and colleagues resorting to frame-by-frame videotape to document the effect).[12]

One thing to note is that while this acquired behavior is itself reflexive and automatic, the appraisal that drives it—an appraisal that involves, for example, assessments of one's relative social position to others—is apparently quite sophisticated, illustrating that even automated tasks can involve sophisticated appraisal conditions. This is important for our purposes, since display rules—rules that govern the proper expression of emotions—are integrally bound up with social categories. For instance, Karla Hoff, Mayuresh Kshetramade, and Ernst Fehr (2011) report that, in India, lower caste groups are less likely to punish norm violations that hurt members of their own caste, in contrast with higher caste groups (controlling for individual wealth, education, and political participation). On the plausible assumption that punishment behavior is subserved by "retributive attitudes," such punishment regularities may reflect complex, category-sensitive display rules.

Display rules not only add to our list of ways that representations differentiate categories, but they provide an example of a way in which "automatic" responses may include not only phylogenetically ancient behavioral dispositions of the sort emphasized by evolutionary psychologists, but also culturally local responses that become "second nature."[13] Automatically conforming to culturally given display rules governing emotional expressions is a shibboleth that displays one's enculturation. Knowing someone has the right sorts of emotional responses is valuable in knowing whether you can coordinate or cooperate with them.

---

[12] This underscores what we saw in Chapter 1: that even phenomena that are relatively culturally invariant can be disrupted.
[13] Though we can ask in turn: is the mechanism that acquires display rules a domain-specific adaptation for acquiring display rule information or a domain-general learning mechanism?

3 Environmental Construction

Our question for now has been: how do social roles produce causally significant categories? Thus far we have answered: by influencing behavior via two psychological paths: by constraining, in virtue of their meaning, rational processes of thought and behavior and by influencing automatic processes. If we return to our simple looping diagram of social construction, we can locate these as components of the top arrow by which our representations determine features of the world that they represent (see Figure 3.1).

So far we have interpreted these constructive processes as involving category representations that influence behaviors towards putative members of categorized groups. However, the model we have created leaves essential mechanisms of construction out. Save for our discussion of acquired automaticity, we have offered little discussion of the role of historical processes and structures in constructing human kinds, nor have we talked about the roles of broader ideological, institutional, cultural, material, and spatial contexts in constructive processes. While these influences are far too multifarious to catalogue in any comprehensive way, they are too important to simply leave out of the picture, and this importance again reveals the shortcomings of focusing narrowly on action and behavior. Ultimately, we need an account that takes these structural elements into account.

Environmental construction refers to the modification of the environment by a person or group that alters the fit of persons or groups to the (now-modified) environment.[14] Common knowledge representations of human categories guide the modification of the environment in ways that systematically affect both members and nonmembers of the categories and also feed back into representational processes. We can depict this basic idea by adding to our simple diagram causal

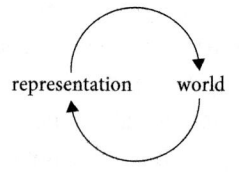

Figure 3.1 Representation-world co-determination

---

[14] "Environmental construction" overlaps with several related concepts: the idea of *niche construction* from biological theory (e.g. Laland et al. 2000; Odling-Smee et al. 2003; Laland and Sterelny 2006), Bourdieu's notion of a *habitus* (1990), and the Heideggerian concept of *throwness* (Heidegger et al. 2010). Each of these related concepts attempts to capture the way that modifications of the environment by one organism or group of organisms produce lasting effects on those that follow.

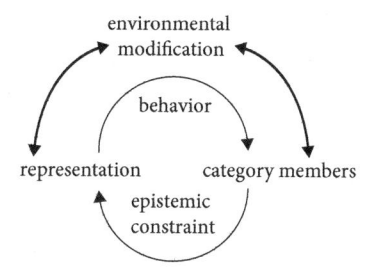

Figure 3.2 Environmental construction

pathways through the environment and noting both their causal effects on categorized persons and on representations (see Figure 3.2).

Changes to features of the environment that are controlled by category representations produce feedback both to the representations that members of the community employ and to the persons that are categorized by those representations.

3.1 Networks and learning scaffolds

A crucial part of our environment is the body of cultures in which we find ourselves from birth. It is a characteristic feature of humans that many of our beliefs and other representational states are acquired from other humans, via social learning. At least in many cultures, they are also acquired from lasting features of our social environment in which they are encoded or inscribed. We in turn select some of these cultural elements to pass on to others, either by directly teaching them, or by further modifying the environment, creating lasting changes that "scaffold" the acquisition of some ideas rather than others (Sterelny 2012). Our existing cognitive representations of human categories shape these projects of acquisition, transmission, and modification by expressing the prior attitudes that guide our decisions about what further information we acquire, transmit, and scaffold.

3.1.1 CULTURAL NETWORKS AND EPISTEMIC WEIGHTING

Cultural networks are sets of people who share information with one another. Since nearly every human is in at least an indirect information-sharing relation to every other, we could regard nearly all humans as comprising a single cultural network. However, if we think of a cultural network contrastively, as *a group of people who share culture with one another more than they do with others*, then we can distinguish networks whose members possess network-specific cultural traits. What sustains a cultural network thus conceived is the mechanism or mechanisms that act as barriers to cultural flow to "out-groups" or amplifiers of cultural flow to "in-groups." And it is plausible that common knowledge beliefs about putatively natural categories figure as such a mechanism, affecting one's choice of cultural "parents" and cultural "offspring." We can elaborate this picture with the idea that we find ourselves not just in a single cultural network, but rather in a system of overlapping cultural networks that influence who we can learn from, what we can learn, and who we can teach.

Miranda Fricker (2007) emphasizes the way socially assigned identities may result in the assignment of different levels of credibility to different speakers. The male African American teen at the coffee shop who is freely offering advice about orthodontic care options may not seem credible, while a middle-aged white male at that same shop saying the same words might seem authoritative. But such assumptions are extraordinarily complex; inverted expectations might apply about another subject: say, local junior high soccer dominance. Such assumptions about credibility are no doubt adaptive and useful in many situations, but as Fricker emphasizes, the effects of systematic influence of this sort affects *who* is allowed to offer testimony and *whose* testimony is accepted and repeated on many matters of importance. Some category members may become "epistemically disenfranchised" while others are epistemically overvalued. Such assumptions about credibility can constitute a mechanism that, in virtue of differentially sorting cultural inputs and outputs, produces cultural networks.

This is just one of possibly many ways that cognitive representations of human categories can structure our cultural networks, determining our possibilities as learners and teachers.

3.1.2 LEARNING SCAFFOLDS

We act as teachers not only by repeating important truths, but by producing long-term environmental modifications that change the possibilities for learning and transmitting—by creating cultural resources that help others determine what to think and how to behave. We can think of these resources formally as theories or learning norms or techniques, but also more materially as the physical encodings of these contents. Consider: books, schools, museums, or public art.

Where such publicly transmitted or scaffolded representational content serves to rationalize existing social arrangements, such modifications may produce and sustain existing differences among category members. Conversely, say, public representations that encode anti-racist or anti-sexist theories may be a standing resource for resisting or reshaping such differentiation. Acting upon and transforming the store of public representations in the cultural environment is thus itself a way of indirectly influencing cognitive category representations and, through them, the construction of categories.

3.2 Institutions, conventions, and norms

Human category representations also figure in structuring the explicit institutions, norms, and conventions that we adopt to regulate our own behavior and that of others. Such elements of social reality regulate behavior in ways that are decoupled from the specific content of what is believed about the category, but that can add to the causal power of the category. Knowing a person belongs to a particular disease category, for example, might allow successful induction regarding disease-typical properties the person has, but it may also allow induction regarding other properties that they come to possess in virtue of the institutional recognition and treatment in a particular community.

3.2.1 INSTITUTIONALLY FIXING ASCRIPTION CONDITIONS

Institutions publicly broadcast ascription conditions for putatively natural categories, allowing common knowledge of category membership to emerge and paving the way for the application of other institutional norms regulating category members. Such ascription takes very different forms for different categories. Consider the case of mental illness as understood in the American Psychiatric Association's Diagnostic and Statistical Manual (the current edition is the DSM-5). A crucial function of this document is to set down explicit criteria for classification that may be used by medical, clinical, financial, and research personnel in inquiring into particular mental conditions, criteria that serve as a "common language to communicate the essential characteristics of mental disorders" for "clinicians and researchers from different orientations (biological, psychodynamic, cognitive, behavioral, interpersonal, family/systems)" (2013, xii). While the manual is explicitly neutral on questions of mechanism, and to a lesser extent even on the ontic unity of the kinds in question, it is ultimately a diagnostic manual that places people into categories so that further, differential treatment (clinical, financial, etc.) can ensue. Here, institutions operate alongside putative natural kinds but nonetheless fix specific ascription conditions for a kind for the purpose of applying further norms, conventions, and institutions.

3.2.2 INSTITUTIONALLY REGULATING CATEGORIES: NORMS AND MATERIAL TRANSFORMATION

Institutions also fix explicit, public norms governing the possible behaviors for members of putatively natural categories. Consider some examples:

- Whether someone counts as a member of a particular Native American tribe may determine whether she can vote for the democratically elected tribal council or whether she can receive a share of a casino's profits.

- For many decades under so-called "Jim Crow" laws in the American South, whether someone counted as white determined whether she was allowed to access numerous privileges including eating in the dining room, riding at the front of the bus, and attending the best public schools. Race-specific restrictions continue in admissions policies to private venues like country clubs and nightclubs.
- Whether someone is a man or woman typically controls what toilets they are permitted to use. Increasingly, universities have been creating both new norms and also organizing their physical facilities to accommodate a range of transgender students and employees.

As norms regulate behavior by and towards members of putatively natural categories, they increase their causal significance.

3.3 Modifying our material environment

Complexly interrelated to the creation of cultural networks and the roles of institutions and norms are other modifications of or systematic effects upon the material environment. By "material environment" here, I draw attention to the artifacts, objects, buildings, land works, and so forth that humans create, and also to the relationship of people to these artifacts and to one another within geographic space. The norms that regulate category members shape access to spatiotemporally located resources, and the management of this access and creation and maintenance of these resources require and are supported by the material transformation of our physical environment, transformations that, in turn, produce their own causal effects on category members.

3.4 Causal significance and covert social role-culture-institution-material complexes

We have been suggesting ways in which our categories guide us as we transform the world, and these transformations, in turn, can add to the inductive power of the categories. I have sketched only a few paths of influence, and real cases will involve exponentially more complex causal networks. Thus, it perhaps is useful to consider real-world cases in which social roles, institutions, and material transformations combine to exert systematic effects.

One important example is the series of Whitehall studies conducted on members of the British Civil Service showing systematic effects of position in the social hierarchy on a range of variables including overall health and coronary heart disease, even when other known risk factors (e.g. smoking rates) are controlled for (Marmot et al. 1978).[15] These studies are so important in part because the subject pool and type of work is alike in so many other ways. They seem to show that mere location in a social hierarchy has systematic effects on human health. Unlike these studies, I have been interested in categories that are putatively natural. Do these categories also exert such influence?

The answer seems to be "yes." Consider, again, race. There are numerous measures on which black Americans have poorer healthcare outcomes than white Americans. While it is difficult to rule out any genetic component in healthcare outcomes, it does seem clear, as Jonathan Michael Kaplan (2010) has emphasized, that social practices of racial classification and discrimination are implicated in producing some of these deleterious outcomes (286f). Kaplan notes, for instance, evidence that native-born black Americans have worse blood pressure than blacks living in other advanced industrialized countries, and they also have worse health outcomes than recent black immigrants to the United States.

We can also see the complexity and causal power of racial classifications reflected in and produced by residential segregation in the United States. Once social groups are physically segregated, a host of other sorts of effects follow, ratcheting up causal differences among category members in sometimes profound ways. Crucially, such cases combine cultural practices of classification, the construction and transmission of cultural resources, the applications of norms, the practices of institutions like governments, banks, and schools, as well as spatial and material transformations to urban spaces.

Many cities in the contemporary United States are segregated by race and ethnicity. This fact has many causes that include:

- historic racial zoning and housing covenants that explicitly restricted residence by race/ethnicity;
- self-segregation by race and ethnicity, including "white flight";
- differential norms of home purchasing and wealth accumulation in different cultural communities;
- the presence of common community resources like schools, subway stops, and parks, which themselves may be differentially distributed.

And segregation also has many consequences, including the production of numerous local generalizations (generalizations about what sort of person lives

---

[15] After the initial study, known as Whitehall I, a subsequent effort (known as Whitehall II) was set up and has now resulted in numerous studies and publications. See the Whitehall Study website for an introduction and list: <http://www.ucl.ac.uk/whitehallII/>.

in what part of town), and the amplification of other racial distinctions (wealth, education, health, cultural) (Anderson 2011).

Because of the range and importance of the resulting effects of geographic racial segregation, Ronald Sundstrom has argued that "attempts to transform social categories must involve the transformation of social space" (Sundstrom 2003, 83), suggesting that past transformation of social space secures the stability of contemporary social categories. On such grounds, Elizabeth Anderson urges the need for integration:

> Segregation of social groups is a principal cause of group inequality. It isolates disadvantaged groups from access to public and private resources, from sources of human and cultural capital, and from the social networks that govern access to jobs, business connections, and political influence. It depresses their ability to accumulate wealth and gain access to credit. It reinforces stereotypes about the disadvantaged and thus causes discrimination. (2011, 2)

3.5 Environmental construction: beyond direct effects of representations

Changes to the corpus of transmitted culture, to lasting institutions, the creation of cultural networks, and lasting changes to material and spatial environments are all products of diachronic processes that produce ongoing effects, amplifying the inductive power of categories around which changes are made. These sorts of causal mechanisms suggest that pure action-based and even behavior-based accounts are insufficient to capture the reality of category construction.

Perhaps it is true, as Arnold Davidson (2001) suggests, that the category of perversion would disappear if our conceptions of sexual instinct also did. If so, this is an indicator that the specific avenue of construction of the category of perversion essentially involves concurrent, ongoing classificatory activities.[16] In contrast, given the numerous environmental ways that racial classifications are encoded in and amplified by environmental arrangements, to arrive late on this scene and suggest that racial difference is merely an institutional consequence of our ongoing conceptual practices with no independent basis seems to elide important facts. Our ideas about race structure not only our treatment of ourselves and others vis-à-vis race, but many other determinants of things that themselves produce racial difference.

---

[16] Other accounts of socially constructed categories have this consequence as well. For instance, if we understand constructed human categories as imposed by the collective imposition of function (Searle 1995, 2010) or as conferred by actual treatment as a member of a category (Ásta 2013), then it seems to follow that the continued existence of the category is grounded in the continued existence of the mental states that allow these actions.

There is, in effect, a continuum for human categories reflecting the stability of their reproduction in the face of various kinds of disruption. Processes of environmental construction stabilize categories in the face of perturbations in the representation of those categories. We can illustrate this idea by considering a philosophers' thought experiment: if we could turn on our Neurosemantic Eraser Ray, blasting an entire urban area so as to erase, say, sexual or racial conceptions and stereotypes from people's brains, producing localized conceptual deficits, to what extent would the categories themselves cease to be? Where a social role is sustained almost exclusively by psychological mechanisms and behavior, the Eraser Ray may be enough to erase the category as well. In contrast, where there is a great deal of environmental construction we should expect that our collective forgetting would be followed by people reconstructing those concepts and stereotypes from stored media repositories of inherited culture, from their own experience, and from the many material facts that encode them. In such a context, concepts would seem to be needed to explain the many distinctions that would continue to co-vary. And public theorizing about such differences might, in turn, give rise to common knowledge of them. If we magically lost the idea of sex or of race, it is plausible that many of us would have to reinvent it in order to understand the social realities that we inhabit.

4 Homeostatic Property-Cluster Kinds

I have surveyed ways that social roles can structure individual behavior on the part of both users of representations and those who fall under the representations, and we have also noted that these social roles also lead to attendant changes to the environment—to the cultural networks, institutions, conventions, norms, spatial arrangements, and material environments that outlast individuals, producing standing circumstances that themselves feed back into the features of individuals. The upshot of all this is that, where there are representations distinguishing category members, the members can come to be increasingly differentiated via causal pathways that include (but are not limited to) intentional actions, representational effects on automatic behaviors, and environmental construction. In this section, I argue that such covert social role categories can be *human kinds*, which is to say, kinds that support induction, prediction, explanation, and intervention in some or another of our inductive projects. I do so by arguing that the kinds amount to *homeostatic property-cluster kinds*.

4.1 The liberalization of natural kinds

Paradigmatic natural kinds for philosophers have been species and elements. But over the last 50 years philosophical work in, for example, the philosophy of mind and psychology has led to relatively wide acceptance of the view that categories from the special sciences that are useful in explanation and prediction count as natural kinds from the point of view of that science (e.g. Fodor 1981, 1997; Griffiths 1999). Richard Boyd (1988, 1992, 1999a) develops an account of the metaphysics of natural kinds that captures the idea that such kinds support our attempts at induction and explanation. Boyd suggests that in our attempts to understand the world we want concepts or terms that pick out causally homeostatic property clusters, the elements of which are co-instantiated in the world. Boyd builds upon this foundation to provide a general explanation of our success in understanding the natural world.

Central to Boyd's account of natural kinds is what he calls *causal homeostasis*: "Either the presence of some of the properties...tends (under appropriate conditions) to favor the presence of the others, or there are underlying mechanisms or processes that tend to maintain the presence of the properties...or both" (1999a, 143). Such a kind is characterized by both,

(1) the properties in the property cluster, and
(2) the mechanism of causal homeostasis—the mechanism that is the source of the properties' continued co-occurrence in the cluster.[17]

As Robert Wilson notes, Boyd's account is a cluster account twice over. First, because an individual does not have to instantiate all the properties in the cluster in order to be a member of the kind (Wilson 1999, 198). (Manx cats do not have tails, but they are cats all the same.) Second, these kinds are cluster kinds because the properties that typify the kind actually 'cluster' in the world. The world is "lumpy" in that these properties are not instantiated evenly throughout space and time, but instead nonaccidentally co-occur.

Consider a paradigmatic philosophical natural kind: water. Instances of water share a variety of superficial properties (e.g. liquidity, freezing point, etc.) and these superficial properties are clustered because they are caused by the microstructural properties of H_2O. The mechanism of H_2O's microstructural properties thus explains the clustering of the superficial properties. This picture of natural kinds famously leads Saul Kripke and Hilary Putnam to view the chemical structure of water as giving its essence (Kripke 1972/1980; Putnam 1975). On this view it is sufficient for some stuff to be water (in any possible world) that it be an instance of H_2O, whether or not it has the superficial properties. And conversely, nothing is water unless it is H_2O. H_2O, on this view, is a property-cluster

---

[17] Boyd has characterized his account several times over the years in ways that are slightly different. See Craver 2009 for discussion of several tensions.

kind in Wilson's second sense only, since instantiating the property of water looks to involve instantiating all the properties that comprise its essence.

Such property clusters support our inductive enterprises because they allow us to draw successful conclusions about other instances of a kind on the basis of examining instances. For example, on the basis of examining very few instances of water, we can successfully infer lots of things about other instances of water. This success is supported by the fact that the properties of water are tightly clustered and causally homeostatic.

But, as Boyd points out, natural kinds in some sciences may be causally homeostatic categories that lack the kind of simple "essence" that chemical compounds like water have. By way of example, he writes,

The appropriateness of any particular biological species for induction and explanation in biology depends upon the imperfectly shared and homeostatically related morphological, physiological and behavioral features which characterize its members. (1991, 142)

In biological species, the instances of a kind at a time instantiate (more or less) property clusters of various sorts of features, but such co-instantiation is imperfect (remember Manx cats). Nonetheless, kind terms picking out species can also figure in successful inductive enterprises. Because species exhibit a variety of properties that are clustered and causally homeostatic, we can—imperfectly—induce facts about other instances of a cluster kind from the few that we actually examine.

Boyd's account is thus a principled liberalization of the idea of natural kinds. This is first, because the property-cluster account of kinds moves from all or nothing accounts of kind membership to a cluster view on which kind membership need not involve satisfying interesting necessary and sufficient conditions. But there is also a second sense in which Boyd's account liberalizes the notion of natural kinds: it expands the sorts of properties that may be contained in the property cluster. While examples like water might lead us to believe that properties that characterize natural kinds or their mechanisms of causal homeostasis must be intrinsic features of members of the kind, reflection on, for example, species suggests that relational properties may be included as well.[18] So for example, according to Ernst Mayr's classic biological species concept, species are groups of populations that "respond to one another as potential mates and seek one another for the purpose of reproduction" (1984, 533).[19] Barriers to genetic

---

[18] The explanatory importance of relational properties has also found many defenders in the philosophy mind, in the debate over broad and narrow content. See Burge 1979, 1986; Jackson and Pettit 1988; Stalnaker 1989; Fodor 1994.
[19] Mayr quite explicitly notes the importance of relational properties to his account, writing "the species is not defined by intrinsic, but by relational properties" (1984, 535).

flow, among other mechanisms, sustain the stability of species property clusters thus conceived. Allowing relational properties to figure as components of causally homeostatic property clusters, or even as the homeostatic mechanisms of such clusters, is a *principled* liberalization of the notion of natural kinds because it is guided by attempts to pick out kinds of explanatory and predictive importance. What is relevant to the ability of the kind to support induction is that there is some stable mechanism or set of mechanisms that causes properties to cluster in a regular way, not whether the properties in the cluster or the mechanisms of causal homeostasis are intrinsic to the kind members. In short: what emerges from Boyd's discussion of causally homeostatic kinds is that properties in a cluster may be imperfectly shared, and also that there need be no a priori restriction on the sorts of properties that may be included in the cluster or the mechanisms of homeostasis, as long as the clusters are doing the inductive and explanatory work required of them.[20]

Opening the door to relational properties in property-cluster kinds also opens up the door to properties that result from human practices, norms, conventions, and so forth. Boyd recognizes as much:

It...follows that there should be kinds and categories whose definitions combine naturalistic and conventional features in quite complex ways...It follows that extensions of the traditional account of natural kinds should be appropriate just to the extent that the kinds in question are employed for induction and explanation. (1991, 140)

It follows that causally significant social roles of the sort we have been discussing could figure as the homeostatic mechanism at the center of important property-cluster kinds that structure our social world. Of course, whether a particular type of social role does constitute an explanatory kind is an empirical question, but social roles that do produce and sustain property-cluster kinds may support induction, prediction, and explanation, earning a place in our best ontology of the social world.

4.2 Entrenched social roles as homeostatic property-cluster kinds

Boyd's homeostatic property-cluster account is often thought of as an effort to make clear how it is possible that a category might support induction, explanation, and prediction in the absence of the sort of simple essence with necessary and sufficient conditions that characterized classic natural kinds (e.g. Griffiths

---

[20] In contrast to some interpretations of homeostatic property-cluster kinds (e.g. Ereshefsky 2007; Bach 2012), I hold that homeostatic property clusters may, but need not, include individuating *historical* properties (cf. Griffiths 1999; Millikan 1999; Wilson et al. 2007). For some arguments against such an allowance, see Ereshefsky 2007, 2010.

1999; Wilson et al. 2007). Because instances of the category share some of the properties in the cluster, and because these properties are reliably co-instantiated in the world, concepts grouping the instances will be useful for explanation and prediction.[21] On this view, the goal of the account is to say, in very general metaphysical terms, what conditions any category would have to satisfy in order to play this explanatory role—to be a causally relevant kind. It seems clear that the entrenched social roles that we have sketched can in principle meet these conditions, for the causal pathways we have emphasized can give rise to and sustain property clusters that would support induction, explanation, and prediction.

5 Inventing Kinds

In this chapter and the last, I have provided a toolbox of some ways in which representing the social world can cause property clusters to emerge and persist. On a category constructionist construal, a category (e.g. race or gender or dissociative identity disorder) may not be a biological kind, but it is not nothing either. It can be a real and important kind structured and sustained by the representations of the category, and by the accumulated effects of such representations.

When information about a putative kind is broadcast by a credible source it can create common knowledge and a social role. Some such social roles, in turn, become entrenched, producing a range of effects that further differentiate putative members of the role. They may begin as somewhat narrow property-cluster kinds centered around a community's labeling practices, and over time grow into more and more substantial property clusters via the sorts of mechanisms discussed here. As these property clusters grow in significance, reference to them may become increasingly important to understanding the social world.

If what I have argued here is correct, category constructionists can offer a range of explanations for the existence and causal efficacy of human kinds.

---

[21] While Boyd's account has come to command a wide following (e.g. Kornblith 1993), recent work on natural kinds has raised a number of problems for it. Craver (2009) notes that distinguishing homeostatic mechanisms may itself be a partially conventional affair, but also that such mechanisms may not really be necessary to a successful account of kinds. Khalidi (2013) and Slater (2015) also suggest that the demand for mechanism need not be met by an account of kinds. Here I take no stand on whether real kinds must always have homeostatic mechanisms, but note that the covertly constructed kinds that we have sketched will for they are just kinds in which the practices structured by the category representation and the effects of these practices produce and sustain difference.

4

Representation and Moral Hazard

> we will say in our tale, yet God in fashioning those of you who are fitted to hold rule mingled gold in their generation, for which reason they are the most precious—but in the helpers silver, and iron and brass in the farmers and other craftsmen.
>
> (Plato, *Republic*, Book III, 415a)

Is addiction a disease, or simply a product of a bad character or weak will? What about obesity? Bulimia? Is sexual infidelity a natural product of our evolutionary design or simply a sign of a bad person or an unjust social arrangement?[1] Questions like these are not merely empirical disagreements over whether a human category or human behavior is a natural kind. They are debates central to contemporary social morality, framing negotiations over which social arrangements, behaviors, and traits are appropriate targets of retributive attitudes, responsibility judgments, and other moral assessments. In discussing them, we automatically sense that there is more than an empirical question at stake; we sense that there are evaluative questions in play.

In a similar vein, longstanding traditions of thought in social theory and in philosophy worry that representing categories of person or behavior as natural may impact both the production of category-related behaviors and also the treatment of those represented. The historian Howard Zinn, for instance, writes in his popular *A People's History of the United States* that,

> This unequal treatment, this developing combination of contempt and oppression, feeling and action, which we call "racism"—was this the result of a "natural" antipathy of white against black? The question is important, not just as a matter of historical accuracy, but because any emphasis on "natural" racism lightens the responsibility of the social system. If racism can't be shown to be natural, then it is the result of certain conditions, and we are impelled to eliminate those conditions. (2005, 30–1)

---

[1] See Stein 1990 for discussion of the complex connection between essentialism about homosexuality and the dispute between voluntarism and determinism (roughly, whether homosexuality is a choice).

Here, Zinn is drawing a connection between the representation of racist behaviors as natural and the moral evaluation of the "social system" or "conditions" that he believes produce them.

Like the social conservative who wants to view homosexual behavior as an individual choice rather than as a manifestation of a natural orientation, Zinn is worried that representing racist attitudes as natural creates a *moral hazard* that threatens to increase the prevalence or prevent the reduction of such behavior. By "moral hazard," I mean a situation in which an actor has reduced incentive to forego a behavior because the actor is protected from some harmful consequence of the behavior.

But why should representation as natural produce moral hazard? I suggest that the answer in these and other cases is the implicit endorsement of something like the following psychological claim:

Reduced Attribution: Representing a human category C as natural reduces attributions of moral responsibility (or related moral evaluations) for instances of C, or for behaviors that are represented as natural consequences of instantiating C. Conversely, representing C as not natural increases attributions of responsibility (or related moral evaluations) for instances of C, or for behaviors that are natural consequences of instantiating C.

Thus understood, Reduced Attribution offers a neat explanation for the production of moral hazard. Insofar as natural explanations lead us to see human actions as, say, less responsible or deserving of punishment, then the costs of those actions will be reduced.[2] Other things equal, this will make these actions more attractive to perform. Reduced Attribution also has considerable philosophical plausibility. Indeed, it may be a consequence of the simple principle *ought implies can*. To the extent that the naturalness of behavior casts doubt on whether it can be changed, it casts doubt on whether a person could be morally required to do so.

Reduced attribution also neatly intersects with social constructionism, for as we have understood them, constructionist explanations are claims that categories or behaviors are dependent upon human decision, culture, and social practices. This intersection is no accident, for reduced attribution plays important roles in covert category constructionist research.

First, *explanatorily*, the idea that representations produce a behavior via moral hazard can itself play an important role in the explanation of the prevalence of a

---

[2] These judgments are not equivalent to one another, but rather exhibit a range that is plausibly implicated as, or as triggering, "reactive attitudes."

behavior. We can see this clearly in Sartre's suggestion that all our putatively natural action is a product of our consciousness, and that the denial of this is itself an attempt to evade responsibility (1956). The same idea emerges in more subtle ways in other constructionist accounts. For example, the constructionist emotion theorist James Averill emphasizes the social importance of the fact that an emotional response "is interpreted as a passion rather than as an action" (Averill 1980a, 312). On his model of emotional behavior, the fact that a behavior is interpreted as a product of nature, of an involuntary "passion," is an important element that facilitates its production. Since representations of a human category as natural can render kind-typical behaviors as seemingly inevitable, they can alter a community's reactive attitudes toward putative kind members and kind-typical behaviors in systematic ways, and plausibly increase the production of kind-typical behaviors.[3]

Second, *politically*, many constructionists hope to replace *exculpating* accounts of behavior as natural or inevitable with those that allow us to hold a person, or a community of persons, responsible for a behavior, thereby prompting social change. For example, Zinn's claim that racism is not natural but a product of social conditions is meant to suggest that we are responsible for our racist attitudes and behaviors and the social conditions that they produce.[4] In both these ways, a belief in Reduced Attribution looks to play a fundamental role in motivating much constructionist research.

More generally, if Reduced Attribution is true, then offering constructionist explanations is a way of engaging questions of contemporary social morality. In this chapter, I develop the idea that Reduced Attribution rationalizes constructionist political concern with the explanation of human categories. Many constructionist theorists are skeptical about scientific claims about human categories, sensitive to the fact that such theories may be distorted by our pre-theoretical assumptions about natural divisions, and suspicious that they may (via a variety of causal pathways) lead to oppressive practical effects. At the extreme, one can view all human representations as merely ideological products of material conditions, or power relations, unresponsive to any independent facts, and therefore

---

[3] To be clear, as we saw in Chapter 3, there are many causal pathways by which the constructionist can connect widespread representations of a behavior with its production, and this is just one of them.

[4] Perhaps the earliest connection between naturalistic theorizing and political order appears in Plato's *Republic*, wherein Plato suggests stabilizing the state by promulgating the "noble lie," as in the epigraph. In contrast with Plato's sanguine attitude towards naturalistic myths, in our own post-Enlightenment age, liberal and critical political philosophers are far more likely to react with alarm at Plato's proposal, guided by concern that accounts of what is "natural" may be used to control or oppress.

debunked by revelation of their contingency. The intimation of such skepticism has led social constructionist work to be viewed by many naturalist critics as head-in-the-sand ignorance, or as so much politically motivated wishful thinking. In contrast, my argument here is that social constructionist concern with representations need not be extreme and is instead rational, given a set of plausible empirical claims. Such concern does not rest upon denying the possibility of successful representation of the natures of human categories. In fact, our practices of formulating representations of human categories carry with them social and political consequences that are obviously of concern.

Here is how I proceed. In Section 1, I consider how best to represent constructionist concern with the representation of human categories or behaviors as natural. I consider accusations by evolutionary psychologists that such concern amounts to the commission of the *naturalistic fallacy*, the fallacy of inferring a normative *ought* from a descriptive *is*. I argue, instead, that such concern is better understood as depending upon the moral hazard produced by reduced attribution of responsibility or related evaluations. In Sections 2 and 3, I review a range of philosophical, evolutionary, and empirical considerations that suggest the truth of Reduced Attribution. I go on, in Section 4, to briefly consider the implications of concern with the consequences of ways of representing.

1 Is Concern with the Content of Explanatory Representations a Fallacy?

Over the last several decades, evolutionary psychologists have put forward a range of sometimes provocative hypotheses about the psychological sources of human behavior, including, for instance, emotional responses (Ekman 1972), human reasoning (Cosmides 1989), cognizing about biological and human kinds (Gelman 2003; Hirschfeld 1996), male and female asymmetries in behavior (Buss et al. 1992; Wilson and Daly 1992). Among the most controversial claims put forward by evolutionary psychologists has been the work by Randy Thornhill and his colleagues (Thornhill and Thornhill 1992; Thornhill and Palmer 2000) arguing that a psychological capacity for rape is an evolved adaptation in human males. Thornhill and colleagues' work has, in turn, prompted substantial critical response, much of it directed at the empirical and theoretical claims that drive their argument.

My concern here though is with a family of responses that is *not* directed at the truth of the theory, but rather emphasizes the deleterious social effects of asserting that a pernicious behavior has roots in an evolutionary adaptation.

Consider, for example, philosophers Kathleen Akins and Mary Windham's (1992) response:

> The hypothesis under consideration is that rape is a "natural" or evolved behavior, one that is "activated" by "environmental cues." Rape is treated, in other words, like any other naturally occurring phenomenon (say lightning); hence the goal of scientific research is construed as the identification of those conditions under which rape "strikes." "If you don't want to be hit by lightning, don't play golf in a thunderstorm"—this, it seems, could be the only practical social implication of the Thornhill view. Rape is explained as a phenomenon that lies outside the realm of moral judgment. (377)

In another response to the same article, the philosopher John Dupré (1992) writes:

> It seems quite clear that the biologicization of rape and the dismissal of social or "moral" factors will both tend to legitimate rape and to deflect attention from social factors such as the depiction of women in the media and advertising and the compliance of the legal system that are plausibly hypothesized to promote sexual violence. Whatever the intent of the authors, their claims will undoubtedly be taken to show that since rape is a "natural" phenomenon, its reduction or elimination is an unrealistic goal. (383)

One mainstay of the evolutionary psychological response to such charges has been to charge its critics with commitment of the "naturalistic fallacy" (Thornhill and Thornhill 1992, 407; Thornhill and Palmer 2000; Pinker 2002, 162ff). The naturalistic fallacy is the fallacy of inferring normative facts from merely descriptive ones—of inferring an *ought* from an *is*. So in reply to responses like Akins and Windham's and Dupré's above, Thornhill and Thornhill (1992) write:

> Rape reflects psychological adaptation and thus the natural processes of historical selection and ontogeny. It does not follow that rape is inevitable or good... Nature simply is, and what ought to be is inferred by people.

Similarly, Steven Pinker writes in response to Dupré, "Note the fallacy: if something is explained with biology, it has been 'legitimated'; if something is shown to be adaptive, it has been 'dignified'" (2002, 162). Evolutionary psychologists argue that no merely descriptive product of a scientific theory can justify normative conclusions, and so suggestions by critics to the contrary are fallacious.

But is that the argument that these critics are making? Let us allow with evolutionary psychologists that the naturalistic fallacy is a fallacy, so that one cannot infer normative facts from merely descriptive premises.[5] Consider the following argument:

---

[5] For a contrary view, see Searle (1964).

Samuel is more than six feet tall.
Therefore, Samuel ought to try out for the high school basketball team.

But no facts about Samuel's height, no purely descriptive facts at all, could possibly allow us to infer what he *ought* to do. To say an argument commits the naturalistic fallacy is to say that it is invalid, that the set of descriptive premises does not entail the normative conclusion.

It is important to emphasize, however, that an argument can be enthymematic—it can assume certain premises without stating them—without committing the naturalistic fallacy. Consider, for instance, this reconstruction of the argument from Zinn:

Premise: Racism is the result of social conditions.
Conclusion: We are morally impelled to eliminate those conditions.

At first look, this might seem a case of the naturalistic fallacy since the premise is descriptive, and the conclusion is a moral imperative. But the argument is really enthymematic, and its validity can be shown by adding in plausible, missing elements.

Premise. Racism is the result of social conditions.

1. Interpreted Premise: If we eliminate social conditions ($c_1, c_2, \ldots c_n$), we eliminate racism (interpretation of Premise).
2. We can eliminate conditions ($c_1, c_2, \ldots c_n$).
3. We can eliminate racism (from 1 and 2).
4. Racism is morally and socially bad.
5. If we can eliminate something that is morally and socially bad, we are (other things equal) morally impelled to do so.
6. We are (other things equal) morally impelled to eliminate racism (3, 4, and 5).
7. Interpreted Conclusion: We are (other things equal) morally impelled to eliminate conditions ($c_1, c_2, \ldots c_n$) (from 1, 2, and 6).

Crucially, these added premises come close to making the resulting argument valid, and indeed while they are disputable, the added premises are at least plausible (as is their attribution to Zinn). In any case, refuting this argument depends not on noting some commission of the naturalistic fallacy, but on more straightforward questions about what is possible, bad, and required.

To be clear: what evolutionary psychologists are objecting to is *not* the production of enthymematic arguments of this sort (see, e.g., Pinker 2002, 164). Indeed, many evolutionary psychologists hope that evolutionary psychology itself might contribute premises to such arguments, supporting novel and successful interventions to solve persistent social problems. Rather, they are objecting to arguments that they think commit a much more straightforward fallacy, something like:

> NFP1. Rape is the result of a natural, adapted mechanism.
> NFC. Therefore, rape is morally permissible.

This argument is indeed fallacious, and it is hard to see what suppressed premises might keep it from being so. The obvious suppressed premise that we might add is simply a commission of the naturalistic fallacy itself—e.g.

> NFP2. If some behavior is a result of a natural, adapted mechanism then it is morally permissible.

Evolutionary psychologists are correct that if this is the argument that critics are making, it is a fallacious one.[6]

However, this is not the best reading of what constructionist critics are doing. The naturalistic fallacy is a red herring. There is a much simpler and more straightforward interpretation of constructionist critics, and it is the one I have suggested connecting Reduced Attribution with moral hazard. In avoiding the naturalistic fallacy, we understand "responsibility" and "moral evaluations" in Reduced Attribution in a naturalistic way: as indicating sorts of thought and behavior such as holding people responsible, blaming people, and punishing people or "giving them what they deserve," and, more generally, with the exercise of what Peter Strawson called exercises of the "reactive attitudes" (1962). Crucially, these are all sorts of judgments and behaviors that can, in principle, be empirically observed and experimentally manipulated. The critics' argument is simply that by promulgating a representation of rape as natural, you *cause it* to be the case that it is taken to be more "legitimate," you *cause it* to be the case that it is taken to be a less appropriate target of responsibility or other moral judgments or reactive attitudes, and thus you *cause it* to be the case that a moral hazard is created that behaviors that are (by agreement) pernicious will be produced and permitted.

Thus understood, the critics' objection is not the fallacious inference from an *ought* to an *is*, but the empirical, psychological claim that representing a behavior as natural causes people to reduce their attributions of responsibility and related attitudes, creating moral hazard, as Dupré himself later makes clear (2001, 89ff). We might represent this part of Dupré's argument as follows:

---

[6] For a different critique of evolutionary psychologists' invocation of the naturalistic fallacy, see Wilson et al. (2003).

D1. Promulgating the view that rape is a product of a psychological adaptation will result in reduced attributions of responsibility and related attitudes.
D2. Reduced attributions of responsibility and related attitudes towards rape will result in more rape, or in less accountability for rape, or both.[7]
D3. We ought not to promulgate that view.

D1 is just the application of Reduced Attribution to this case.

Reduced Attribution thus rationalizes constructionist concern with the content of representations of human categories without implying any global skepticism about empirical knowledge of human nature and without commission of the naturalistic fallacy. And crucially, as we just noted, Reduced Attribution is both plausible and testable in both general cases and in this specific one.

2 Arguing for Reduced Attribution

Suppose this reading of constructionist concern is correct, and suppose that Reduced Attribution is appropriately understood as an empirical claim. There remains the question of whether Reduced Attribution is actually true. While evidence on this question is still emerging, in this section and the next, I argue that it is plausibly true. I begin by considering a conceptual objection to Reduced Attribution. I then consider some philosophical and evolutionary considerations that also suggest its truth before turning in the next section to review a range of experimental evidence for Reduced Attribution.

2.1 Representing as natural

We have already noted the intersection of Reduced Attribution's use of the category of *natural* with constructionist explanations that oppose natural explanations in favor of those that appeal to human decision, human culture, and social practices, and we suggested that this intersection is illuminating because Reduced Attribution thus provides one justification for why constructionists are concerned with the category of the natural. But this opposition to natural explanations and this reliance on an empirical assumption that natural explanations tend to be exculpating might be thought to be bad news for constructionists for it is not clear that "natural" can be given any precise content. Thus we might also wonder if there is any good reason to think that the empirical assumption is true.

Begin with the thought that, while it may seem intuitive to contrast natural behavior with that which is the result of "decision" or of "nurture" or "culture" or

---

[7] See Dupré (2001, 90-1) for his development of something like Reduced Attribution.

"social construction," it is not at all clear that this intuitive contrast can do any further work since it is implausible that, say, human decision or human culture or social practices are in any metaphysically substantive sense nonnatural.[8] While there is a long tradition of humanistic and interpretive social science (from which some social constructionist work grows) that suggests cultural and psychological explanations stand outside of the causal or explanatory order revealed by natural science, this view is very controversial. Adopting it would mean abandoning large swaths of the human sciences that now exist. In any case, it is part of my project in this book to naturalize social constructionist claims, showing how they can be understood to cohere with what the sciences tell us about the world. Looked at in this way, social constructionist explanations, like the "natural" (e.g. nativist or biomedical or genetic) explanations they often contrast with, are also causal or constitutive explanations that depict culturally produced behaviors as part of the causal order. If, as philosophical naturalists (I think rightly) assume, everything—or at least everything short of perhaps a few things (maybe sets or qualia or moral value)—is natural in this sense, then the distinction at the core of constructionist work seems to be in trouble.

While constructionist explanations, and Reduced Attribution, may imply a contrast between what is natural and what is not, the distinction they are interested in (for both explanatory and political purposes) does not actually depend on a metaphysically deep distinction between the natural and nonnatural. It depends instead only on a far more plausible assumption that certain descriptions of traits or behaviors (the ones that we represent as "natural" ones) are treated as exculpating while others (the "nonnatural" ones) are not. Thus understood, the truth of Reduced Attribution depends upon the existence of a psychological asymmetry in the way various sorts of explanations are treated by the mind. While there is work to be done to discern the exact boundaries of this asymmetry, there is growing evidence of the existence of such an asymmetry, evidence that I turn to momentarily. For now, the important point is that the category of the "natural" that figures in Reduced Attribution, and whose complement is employed in our account of constructionist explanation, is itself a category that figures as half of this asymmetry. Its precise borders will be revealed in the course of further empirical investigation, but for present purposes we can characterize it as the category of things that are not regarded as appropriate targets of moral attitudes like blame, holding responsible, and punishment.

---

[8] For more general considerations to this conclusion, see Chapter 6.

2.2 Philosophical roots of reduced attribution

Understanding this sort of thought and behavior as sensitive to the way we represent an agent vis-à-vis a sort of behavior is a familiar philosophical theme. Influentially, Strawson's (1962) suggestion was that when we take the "objective attitude" towards a human (treating a human as an object rather than a person), it tends to undermine the application of the reactive attitudes—attitudes like anger, praise, blame, holding responsible, and so forth. And categorizing a person's behavior as a product of a natural category may be a way of taking such an objective attitude. Implicit here is that the exercise of the reactive attitudes is not "all or nothing." Rather, it is graded and modified by our beliefs about the sorts of natural categories to which a person belongs, about the nature of those categories, and about their "quality of will."[9]

This theme also emerges in Sartre's early work. Central to Sartre's early thinking is the idea that we are essentially free and responsible agents, and that our actions, even our putatively natural behaviors, are under the control of our conscious will. On Sartre's view, it is always a mistake to take an objective attitude toward others or ourselves; doing so is a way of allowing or attempting the evasion of responsibility. Sartre's examples register especially the ways in which putatively natural categories of behavior can play such an exculpating role. For instance, he considers apparently natural emotional responses, writing of sadness:

> What is the sadness, however, if not the intentional unity which comes to reassemble and animate the totality of my conduct? It is the meaning of this dull look with which I view the world, of my bowed shoulders, of my lowered head, of the listlessness in my whole body... Is not this sadness itself a *conduct*? Is it not consciousness which affects itself with sadness as a magical recourse against a situation too urgent? And in this case even, should we not say that being sad means first to make oneself sad? (1956, 104)

Sartre's view has it that the putatively natural order exhibited in our thought and behavior results from the intentional activity of organizing our behavior around a representation of a category in order to achieve benefits or avoid sanction, and our failing to acknowledge this amounts to bad faith. We can acknowledge Sartre's contribution without endorsing his radical break with the objective attitude. Like Strawson, Sartre recognized the role that certain sorts of representations could play in both the production of behaviors represented as natural and

---

[9] David Shoemaker (2015) develops a Strawsonian account in which he focuses on what cases of marginal agency can teach us about the (for Shoemaker, "qualities of") will and their relationship to responsibility.

in regulating responsibility attributions toward oneself and others. He believed in the truth of something very like Reduced Attribution.

2.3 Evolutionary conjectures about reduced attribution

Why should Reduced Attribution be true? Looked at in an evolutionary context, our reactive attitudes are part of a suite of responses that support human ultra-sociality, responses that allow us to live in enormous groups of other humans, our behavior successfully regulated by relatively abstract moral norms. Plausibly, such norms themselves are sustained by practices of punishing norm-violators, and perhaps by practices of meta-punishment (punishing those who fail to punish) of various orders.[10]

If function of our reactive attitudes is in substantial part to regulate our own and others' behavior, it makes sense that we do not grow angry at the regular movement of the stars across the night sky. Nor do we attempt to hold the moon responsible for its waxing and waning. Nor do we view the tree as culpable for the changing color of its leaves in autumn and their loss in winter.[11] Exercising our reactive attitudes comes at cost to us, and when we exercise them in response to phenomena that are not themselves sensitive to our attitudes and behaviors, then they fail to regulate the target phenomena and waste our resources. In contrast, when we exercise our reactive attitudes in response to persons whose behavior is sensitive to the attitudes and actions of others, these attitudes can play a regulative role, altering the behavior of the target of our attitudes and signaling to others some important feature of the social environment (for example, that someone is not cooperating).

3 Psychological Evidence for Reduced Attribution and Moral Hazard

I have offered anecdotal examples, and philosophical and evolutionary theories, to support the truth of Reduced Attribution, but what remains is to document the phenomena with experimental work. In this section, I consider a range of work that supports Reduced Attribution and its connection with moral hazard.

---

[10] There is now an enormous literature on the evolution of norm-governed behavior. Some useful starting places for exploring these issues are Richerson and Boyd (2005); Bowles and Gintis (2011); Kitcher (2011); Sterelny (2012).

[11] Specific judgments about what counts as natural in this sense may vary considerably from culture to culture.

3.1 Uncontrollable causes result in reduced attribution

I begin with experimental results of moral judgment that have connected judgments of the control over a behavior with blame for it. In an early study, John Monahan and Gloria Hood (1976) described a violent act as the product of a psychiatric disorder, and found significantly reduced judgments of both voluntariness and blameworthiness. Similarly, Bernard Weiner, Raymond Perry, and Jamie Magnussen (1988) found that describing a trait like *blindness* or a behavioral pattern like *being a child abuser* as the product of accidental or uncontrollable origins reduced judgments of culpability for the condition as compared to an origin under greater perceived control. More recently, John Monterosso, Edward Royzman, and Barry Schwartz (2005) found that describing behaviors as resulting from physiological causes (neurotransmitters) significantly lowered culpability assignments across a range of vignettes when compared to causes rooted in experience (e.g. abused as a child).

These judgments of reduced culpability plausibly translate into reduced punishment. Azim Shariff and colleagues carried out a series of experiments showing that representations of the neural bases of human behavior reduced support for retributive punishment when subjects were asked to act as hypothetical jurors (2014, experiments 2–4). Relatedly, a recent study of actual judges suggested reduced sentences for incurable psychopaths who committed aggravated battery when their behavior was explained as a product of genetic and neurobiological factors (Aspinwall et al. 2012).

While these are complex experiments, it is plausible to think that the upshot of them is that culpability and punishment for a trait or behavior varies directly with the level of control that an agent is perceived to have over it. Insofar as behaviors or traits represented as involving "natural" causes suggest less control, agents will have reduced culpability. This is evidence for the truth of Reduced Attribution.

A burgeoning literature in experimental philosophy of free will also seems to support Reduced Attribution and suggests an explanation for it. Eddy Nahmias and his collaborators (Nahmias et al. 2007; Murray and Nahmias 2014) have argued that exculpatory judgments emerge from causal explanations of behavior on which the processes giving rise to the behavior seem to bypass processes of executive control. If certain representations of behavior implicate less calculation or choice on the part of the agent, they may imply less responsibility. Another possible explanation is that the "control" processes involved in calculation or choice matter because they indicate an ability to do otherwise. Robert Woolfolk, John Doris, and John Darley (2006), for instance, have found that constraint of an agent of various kinds resulted in decreased judgments of responsibility across a range of conditions.

3.2 Reduced attribution and moral hazard

Constructionist concern emerges from drawing a connection between Reduced Attribution and moral hazard. So another question is whether a reduction in the attribution of moral responsibility actually leads to moral hazard? Reduced attribution of responsibility ought to produce moral hazard to the extent that humans are rational. Other things equal, there ought to be an increased propensity to produce actions as the costs of the action are reduced or as the benefits are increased given our approximately rational nature. Thus understood, social constructionists are (along with much of the social sciences) engaged in a form of rational choice explanation.

These rational choice assumptions are enormously plausible in this context. I began this chapter by considering debates in contemporary social morality over the "naturalness" of certain behaviors or categories, and the existence of such debates already reflects widespread if often implicit recognition that how we explain actions is connected to the appropriateness of a range of moral attitudes towards the actions. We can see such recognition manifest in, for instance, the finding that whether a behavior is explained by appeal to a feature of the person or the situation seems to depend on whether the event is positive or negative (suggesting a self-serving motivation to use ascription as a means to negotiate praise and blame) (Malle 2006).

More direct evidence for moral hazard can be found as well. For instance, Crystal Hoyt, Jeni Burnette, and Lisa Auster-Gussman (2014) placed obese subjects in a condition in which obesity was represented as a disease (as the American Medical Association indicates), and these subjects subsequently showed reduced importance placed on health-focused dieting and made higher-calorie food choices. Interestingly, they also showed reduced concerns about weight and lowered body-image dissatisfaction.

Although their study does not show this, one explanation of Hoyt and colleagues' results appeals to the inferences people draw from the assertion that obesity is a disease. For example, obese subjects might infer from obesity's status as a disease that their own actions will be ineffectual in fighting obesity. Or they might infer that interventions cannot work to correct obesity. Or perhaps they infer that obesity is genetically caused and so they are not to blame for being obese. These inferences themselves may be unsound, but the crucial thing is that, insofar as subjects make them, representations may result in a reduction of attribution of responsibility, producing moral hazard.

Recent work in the experimental philosophy of free will also suggests the production of moral hazard. In one prominent study, Kathleen Vohs and Jonathan Schooler (2008) found that subjects who read a description of the brain that denied the existence of free will were more likely to cheat on a subsequent task than those who read a different passage.[12] Following on this study, Roy Baumeister, E. J. Masicampo, and C. Nathan DeWall (2009) found that inducing disbelief in free will reduced helping behavior and increased aggression, and chronic disbelief in free will was also connected with reduced helping behavior. In both studies, belief that one's actions are causally determined seems to create moral hazard to engage in anti-social behavior. Plausibly, this occurs because Reduced Attribution is true: one sees oneself as naturally, causally determined, and infers from this that one is not an appropriate target of moral judgment by oneself or others.

This emerging experimental evidence points in the same direction as the previous considerations, namely to the conclusion that Reduced Attribution is true and can produce moral hazard.

4 Vindicating Constructionist Concern: Beyond Truth

I have developed an interpretation of concern with human category representations as growing out of an appreciation of Reduced Attribution and a concern with moral hazard, and I have argued that such concern is empirically plausible. Such concern is not a reflection of skepticism about the possibility of scientific knowledge of human categories or behaviors; in fact, it depends in part on empirical evidence linking certain sorts of representations to Reduced Attribution.

But there is something of fundamental concern to philosophers and scientists at stake in constructionist concern with representation, and with constructionist arguments involving Reduced Attribution: the permission to promulgate hypotheses about human behavior that may themselves be not what we hoped for, or claims that are themselves capable of being put to repugnant or pernicious use. It seems plausible that many scientific theories of human categories can, via Reduced Attribution and its connection with moral hazard, sometimes promote

---

[12] A subsequent attempt to replicate the first experiment in this paper found a much smaller effect on cheating that did not reach significance, calling into question the original result. In addition, an analytic mistake in analyzing the second experiment led to a smaller effect size than originally reported (though still significant) (Giner-Sorolla et al. 2015). Despite this, the overall evidence (of these and related studies) seems to continue to point in the direction of the creation of moral hazard.

harmful behaviors. Does that mean such theories should not be expressed or entertained in the humanities or the human sciences?[13]

That view is far too strong. For one thing, it probably applies to nearly any theory of human nature or behavior, undermining the possibility of human science altogether. For another, it thwarts possible interventions that might ameliorate the harmful effects of certain traits. If—as I argued in Chapter 1—racial essentialism really is a manifestation of an innate human tendency to cognize about certain topics in an essentialist way, this insight may lead to new ideas about how to undermine these erroneous essentialist thought patterns. For instance, elsewhere I have suggested that the failure of such essentialist mechanisms to determine categorization in mixed-race cases lends support to a suggestion from Naomi Zack (1993) that the ubiquity and complexity of mixed-race cases undermines racial categorization (Mallon 2010; Mallon and Kelly 2012). More generally, if we read Zinn's opposition to natural explanations of racism as restricting research into the psychology of racism, we would miss out on potentially useful social interventions (cf. Kelly et al. 2010; Machery et al. 2010).[14] Prior constraints on the acceptable contents of scientific hypotheses prejudge important questions, leaving us without possibilities for insight and intervention that we would otherwise have had.

A different thought is that we should not express or entertain theories publicly, or in specific public contexts, unless the theory meets some threshold of probability. In responding to Thornhill and Thornhill (1992), John Dupré suggests such a view when he writes, "because, as far as I can see, the Thornhills say nothing that even affects the probability of their truth, the propagation of such claims should be strongly resisted" (383; cf. 2001, 89ff). And sensitivity to the potential impact of a theory seems relevant to establishing the threshold of probability required to assert its truth. Relatedly, Cordelia Fine has suggested that the standards of empirical evidence ought to be higher for asserting neurobiological gender difference, given the possibility that the promulgation of such assertions may have harmful social consequences (2010, 174). To a large extent, requiring probability in advance of promulgation is what the gatekeepers of professional status—conference organizers, journals, publishers, and, perhaps

---

[13] It is not only constructionists who critique opponents' views in terms of potentially harmful consequences. For instance, philosopher of biology David Buller (2005a, 2005b) has critiqued Martin Daly and Margo Wilson's evidence for a greater rate of mistreatment of children by stepparents than genetically related parents. Daly and Wilson (2005) replied (in part) that "distorting what is known about family violence for rhetorical purposes could do real harm in the practical realm of child protection" (508).

[14] In the quoted passage, Zinn is not recommending restrictions on speech, but rather framing an empirical, historical argument against the naturalness of racism.

journalists—are supposed to do. But exactly what the threshold of probability for various claims is, and whether it is better to lower it in order to encourage provocative debate and research or raise it to discourage potentially harmful speculation, are subjects inviting sharp disagreement driven, in part, by competing values.

In any case, finding some threshold of probability for a theory to be discussed or taken seriously or promulgated does not resolve the concern. We can see this by considering that utterances can have harmful consequences *even when justified or true*. Recall, for instance, Hoyt and colleagues' evidence that moral hazard is produced by describing obesity as a disease (which it is, according to the American Medical Association). Even if obesity is a disease, labeling it as such in some contexts may have detrimental consequences because of the psychological propensities of the consumers of those representations. Similarly, Shariff and colleagues (2014) manipulated subjects' judgments about punishment simply by providing information about the neural bases of behavior (experiments 3-4). Again, the representations involved are plausibly justified and may be true; their promulgation may also be harmful.

The sorts of consequences suggested by Reduced Attribution may offer nonepistemic grounds on which to object to utterances, grounds that may be considered alongside epistemic values. It is easy to generate hypothetical cases in which nonepistemic considerations offer overriding reasons to object to utterances that might be epistemically justified or true, or to make utterances that are false. When the mad dictator threatens to blow up the world unless you deny that 3 is a prime number, you should do so. When the Nazis are at the door, it's okay—even morally required—to lie to them about who is in the house and to object to those that do not. Once we recognize such nonepistemic considerations, and the possibility that they can be overriding, it follows that conceding the truth or high probability of a claim does not solve questions about when and where the claim ought to be asserted.[15] Uttering and asserting are actions, and they are subject to evaluation as actions along a wide range of dimensions.

But while an exclusive focus on epistemic evaluation leaves out nonepistemic concerns, so nonepistemic concerns "change the subject" for many, and perhaps most, practitioners of philosophic or scientific inquiry. For discourses aimed at truth, nonepistemic considerations seem to be the "wrong kinds of reasons"

---

[15] Sally Haslanger has recently explored the possibility of social critique of claims that are nonetheless justified and true but that can be critiqued, in part because the statements play a role in constituting the social practices that make them true (2012, 407).

to consider.[16] How can we accommodate respect for assessment in terms of justification, probability, and truth and the consequentialist assessment of a discourse? This question probably has no general answer. In some cases, it can seem obvious that one or another value is overriding, but resolution plausibly involves real conflicts and tradeoffs where respecting one value means ceding satisfaction of another.[17]

Throughout this book and in this chapter, I employ many social scientific representations of human categories in order to develop a sympathetic account of constructionist approaches to human categories. As this suggests, I do not hold that we should abandon our attempts to scientifically represent human categories and human kinds, nor should we be cowed by what are often poorly understood consequences. But as we represent, we engage a complex causal network of mental states and dispositions that itself includes standing propensities to think and behave in certain ways. These propensities sometimes give empirical support to concerns about the consequences of representing human categories in certain ways, even where the representations are justified or true, concerns that deserve to be addressed.

It is not surprising that different theorists with different assignments of probabilities and different values come to disagree about what should or should not be asserted or promulgated regarding a domain. The sort of evidence considered in this chapter does offer one possible, general avenue for making progress on these difficult evaluative questions. Careful empirical investigation regarding our cognitive proclivities to take certain representations as triggers for further misrepresentations may help us better understand the risks and rewards of various approaches to stating and promulgating our theories.

---

[16] The phrase the "wrong kind of reason" is typically used in discussions of reasons for values, but here (following Pamela Hieronymi), I use the phrase to discuss reasons for *beliefs*. As Hieronymi observes, "there is a quite general problem about identifying the appropriate reasons for attitudes, a problem that is not restricted to reasons for those attitudes involved in valuing" (2005, 438 fn. 2).

[17] See Stein 1999, chapter 12 for discussion of the various arguments for and against research into whether scientific research on sexual orientation should be carried out.

5

Performance, Self-Explanation, and Agency

I have been developing the idea that social constructionist explanations of human thought and behavior hold that our representations of ourselves, as humans or as members of various specific categories of human, produce and regulate the categories, thoughts, and behaviors of those they represent. One provocative and influential variety of social constructionist account is a performative constructionist account. Performative accounts explain thoughts and behaviors as products of an intentional, strategic performance elicited and regulated by our representations of ourselves, with our representations sometimes eliciting the very thoughts and behaviors they represent.

I have already mentioned a number of (at least partially) performative accounts, including Ian Hacking's fascinating accounts of "making up people" (1986, 1992b, 1995a), as well as K. Anthony Appiah's work on racial identity (1996, 2005). Similar accounts have figured in constructionist accounts of the emotions (Averill 1980a, 1980b; cf. Griffiths 1997) and anti-psychiatry accounts of mental illness (e.g. Scheff 1974; Goffman 1970). And they are implicit in Sartre's (1956) discussions of freedom and bad faith. Their presence in contemporary intellectual life is perhaps most keenly and persistently felt in feminist philosophy, where performative accounts offer an explanation of gender differences, especially behavioral differences (Butler 1990; MacKinnon 1987).

While these performative accounts are widespread and influential, they receive little discussion in philosophy of mind and psychology. Perhaps this is because the accounts themselves seem implausible or because they concern culturally shaped categories that are too ephemeral to be worthy of sustained study by psychological inquiries focused on more fundamental mechanisms. Or perhaps the view is that such accounts are simply too shaped by antecedent political sentiment to be seriously considered as empirical hypotheses. Whatever the cause, the result is a disciplinary gulf in which some explain category-typical human behaviors by appeal to intentional responses to the symbolic practices surrounding the category, and others offer little engagement of these explanations.

I have four aims in this chapter. First, I want to use the account of entrenched social roles offered in Chapters 2 and 3 to offer a causal model of performative social constructionism. A second aim is to clearly articulate a puzzling feature of performative claims that makes them seem especially implausible, one I call *the puzzle of intention and ignorance*. The puzzle results from two strands of performative constructionist theorizing. One strand is simply the performative explanatory strategy: performative constructionists try to explain category-typical thoughts and behaviors, for some category C, by appeal to intentional action. The second strand is the Revelatory Aim of constructionism mentioned in the Introduction. Like many social constructionist explanations, performative constructionists are especially interested in explaining thoughts and behaviors that are widely but mistakenly believed to be the unintentional result of nature—perhaps the consequence of membership in a natural kind. So, for example, performative gender theorists seem to hold both that gendered behavior is intentionally enacted, but also that gendered behaviors are widely and mistakenly believed to be products of a natural kind. The puzzle is: why doesn't the intention undermine the ignorance? If we are all, say, performing gender difference intentionally, how could we be so mistaken about this? This is the puzzle of intention and ignorance.

My third aim is to resolve this puzzle in a psychologically plausible way. Across a number of fields, a common way to explain intentional but misexplained behavior is by appeal to failures of self-knowledge. But how does such a failure occur? I suggest that a plausible understanding can be found in the failure to locate one's mental states in a causal explanation of one's thoughts and actions. This understanding, when added to the social role account developed in Chapters 2 and 3, completes a sketch of a causal model of specifically performative constructionist claims.

With this causal model in hand, my fourth aim is to use it to illuminate a connection between how we theorize about human categories, and the moral agency of humans that fall in those categories. Specifically, I argue that the sorts of theories we (as a community or as a culture) offer of particular behaviors can create or destroy agency with regard to those behaviors. This conclusion, in turn, adds to the discussion of Chapter 4, further illuminating and rationalizing persistent concern within social and political theory that how we represent human nature or category-specific natures may undermine human freedom in nonobvious ways.

Here is how I proceed. In Section 1, I review some examples (some now familiar) of accounts of human categories that explain category-typical behavior, at least in substantial part, as a performance of the representation of the category. Then, in Sections 2, 3, and 4, I go on to develop a causal model of such performativity. Specifically, in Section 2, I briefly articulate the application of the category constructionist social role model developed in Chapters 2 and 3 to the performative case. Then, in Section 3, I turn to set out and motivate the problem of intention and ignorance for performative constructionists and suggest a solution in failures of self-knowledge. I go on in Section 4 to articulate a constructionist account of failures of self-knowledge involving a failure to locate. Section 4 thus supplements the constructionist social role model with a psychological model of the mechanisms that might underlie performativity (answering the problem raised in Section 3). In Section 5, I connect the model I have developed with agency and responsibility, arguing that how we represent the categories to which we belong, and therefore explain the category-typical behaviors that we perform can constitute or fail to constitute us as agents with regard to those behaviors. I conclude in Section 6.

1 Explaining Behavior as a Performance

While only a few theorists use the terms "performance" or "performative" to describe their explanations of behavior, many more offer what I call performative explanations that share the following two properties:

(1) The account explains category-typical thoughts or behaviors of putative category members by appeal to these members intentionally acting so that their thoughts or behaviors realize (or appear to realize) properties represented in a theory of the human category. In effect, the theory of a category forms a guide for improvising thoughts and behaviors.
(2) The thoughts or behaviors are widely believed to be (in some way not always specified) the natural result of membership in a category.

Malingering provides a clear illustration of this dynamic. In malingering, there is a widespread understanding of a category that is believed to be natural and involuntary. A person may then enact symptomatic behaviors of category members in order to achieve benefits of some sort from others. These benefits accrue to the malingerer precisely because of a propensity by others to explain the behaviors (e.g. an inability to work) as the natural and involuntary result of membership in the disease category.

However, the accounts we consider below differ from malingering in three crucial respects. First, unlike imitating a natural illness category, constructionists about a category often insist that there simply is no natural category, and that putative instances of the category are all products of our representational practices. For example, many theorists believe race and gender are not natural kinds while allowing that our race and gender classifications track real differences in the population. Social constructionists explain these differences as the result of our classificatory practices, and the performative cultural constructionist offers one version of this explanatory strategy—a version that explains differences as the result of intentionally performed action. Second, malingering involves lying. But many performative constructionist accounts aim to explain thoughts or behaviors about which the actors themselves may sincerely offer mistaken explanations. I explore how this is possible in Sections 3 and 4. Third, we think of a malingerer as cheating to get benefits, perhaps by shifting burdens to others. But occupants of social roles should be thought of as responding to incentives in a much broader way that includes not just seeking benefits by shifting burdens, but also such things as escaping coercion and violence, resolving social ambiguity, achieving moral license and exculpation, cementing or dissolving social alliances, avoiding and coping with aversive emotions, and so forth.

As we have noted, performative accounts have been applied to a wide range of categories, from the exotic to the everyday. Here, I rehearse a few examples.

1.1 Making up people

Recall from Chapters 2 and 3, Hacking's exploration of ways in which changing classifications of persons lead to new ways for a person to be, including his extended (1995a) examination of multiple personality disorder. Crucially, Hacking holds that multiple personality disorder is a performance in the sense described above. On his view, the representation of the category guides the actor towards various social benefits. Recall Hacking's remarks that:

> When new descriptions become available, when they come into circulation, or even when they become the sorts of things that it is all right to say, to think, then there are new things to choose to do... Multiple personality provided a new way to be an unhappy person. Even many supporters of the multiple personality diagnosis are willing to agree that it has become, to use one popular phrasing, a culturally sanctioned way of expressing distress.
> (Hacking 1995a, 236)

This sort of account explains (at least a significant component of) multiple personality behavior as intentional action.[1]

---

[1] As I noted in Chapter 2 (section 2.2., fn. 4), Hacking expresses doubts about the sufficiency of conscious, intentional action to account for the phenomena of multiple personality disorder or other

1.2 The performance of everyday categories

While performative accounts may have plausibility when applied to an unusual category like multiple personality disorder, can they work for more ordinary but socially contentious categories like race, gender, sexual preference, and the like? In Chapter 4, we saw that Sartre, an influential precursor to performative constructionists, says "yes" (Sartre 1956). Central to Sartre's early thinking is that we are essentially free and responsible agents, and that our actions, even our putatively naturally caused behaviors, are under the control of our conscious will. (Recall the example of sadness as "a conduct.") Sartre's view has it that the putatively natural order in our thought and behavior results from the intentional activity of organizing our behavior around a representation of a category in order to achieve social benefits or avoid social sanction, and failing to recognize this amounts to bad faith. Thus, Sartre's view amounts to an assertion of a ubiquitous role for intentional, strategic action in cases of behavior produced by putatively natural categories like the emotions.

We also saw (in Chapter 3, Section 1) Appiah's suggestion that racial identities (among other social identities) are produced and controlled by culturally local understandings with real behavioral effects. Appiah writes that,

> [Racial labels] shape the ways people conceive of themselves and their projects. In particular, the labels can operate to shape what I want to call "identification": the process through which an individual intentionally shapes her projects—including her plans for her own life and her conception of the good—by reference to available labels, available identities. (1996, 78)

While Appiah holds that the applicability of racial labels is not a matter to which we consent, he (like Sartre) holds conforming our thought and behavior to such labels is ultimately a matter of choice, writing that for a racial identity "I can choose how central my identification with it will be—choose, that is, how much I will organize my life around that identity" (1996, 80).

And we also saw Judith Butler's famous claim in *Gender Trouble* that gender "has no ontological status apart from the various acts which constitute its reality," but that "words, acts and gestures, articulated and enacted desires create the illusion of an interior and organizing gender core" (1990, 136). Here, as in Appiah's description of racial identity, the idea of a putatively natural sort of person structures performances—performances that may enact representations and simultaneously sustain the illusion of their naturalness.

"interactive kinds" (e.g. Hacking 1995b, 368, 1999, 115). The discussion here assumes only that performative accounts treat it as necessary.

1.3 Target claims

The range of performative claims alluded to in this section amount to my target claims in what follows. In each case, the causal mechanisms at play are somewhat underdescribed. But, here again, in offering a psychological model of such performances, my aim is to offer an account on which, if the model were realized for the relevant category, the target performance claims could turn out to be true.

2 Social Roles and Performance

Communities use classifications of persons as a basis for prediction, explanation, and coordination, and this creates the social context in which individuals in the category develop, live, and act. Social constructionists (of various stripes) about human categories then explain the existence or differential features of the category (or the significant features, or some significant features) by appeal to our collective practices of thinking about and acting toward putative members of the category and occupants' responses to these. On the account I offered in Chapters 2 and 3, such practices involve category representations that express *common knowledge*—beliefs that everyone has, and everyone knows that everyone has (and so forth)—for such common knowledge provides the shared understanding that is the basis for coordination and cooperation with regard to category members. It also provides the basis for performance of category features by role occupants. Where there are social roles of this sort, those who occupy the social role (by meeting the conditions implicit in the conception of the role) will thereby have a range of properties as a result.

Here, our target claims are concerned especially with the explanation of thoughts, behaviors, and dispositions to think and behave of the persons who fall under (or are believed to fall under) a classification. And even brief consideration of the effects of social classification will suggest that there are many, many ways that occupying a social role may affect the thoughts and behaviors of a person. For present purposes, we distinguish these causal pathways according to whether or not they are *proximally intentional*.

We can treat a thought, disposition, or action as intentional just in case it results in the normal manner from a thought process in which a thinker/actor's beliefs and desires interact in a rational, "reasons responsive" way.[2] An action or thought is proximally intentional if it is the proximal or immediate result of such

---

[2] The intentional domain may be much broader than this, encompassing behaviors that are not "normal" or "rational." Certainly, we attribute intention in some such cases (e.g. cases of akrasia). For present purposes, I am not concerned with this broader domain.

mental states. For example, if Etta wants to go to the cinema, checks the listings and her watch, and hops on the tram to catch a 7:15 film, she is engaged in ordinary intentional activity—activity in which her beliefs, desires, intentions, and actions give rise to further beliefs, desires, intentions, and actions according to norms of theoretical and practical rationality. An important feature of such proximally intentional states is that we are often conscious of them, both as they happen and later on, as we engage in self-explanation.

In contrast, a disposition to think or behave is antecedently intentional if it results from a process or mechanism that is itself produced by past intentional activity on the part of the actor. With practice, some activities that begin as the results of conscious intentional action become (in full or in part) automatic (or unconsciously carried out) over time, and this applies even to some cognitively demanding tasks, like recognizing good chess moves.[3]

Still other behavioral dispositions may not be mediated by a subject's intentional states at all. For example, they may be due to a long process of externally produced behavioral reinforcement. Or they may be the result of innate, automatic mechanisms—for example, functionally specific cognitive modules (e.g. Tooby and Cosmides 1992).

With these distinctions in mind, we can now note that the really distinctive thing about performative social constructionist claims is that they seem to suggest that to some significant extent putatively natural category-typical dispositions and behaviors are, or result from, proximal intentional activity on the part of people occupying a social role (Griffiths 1997, 149ff).

On the one hand, that a social role affects proximal intentional behavior is not at all surprising. The explicit and implicit conventions, understandings, and norms that surround each of us offer social affordances including restrictions, permissions, obligations, as well as opportunities for self-presentation within which we can rationally and strategically act to advance our aims. But on the other hand, this sort of view is particularly surprising in that for many objects of performative constructionist claims, constructionists pair a claim of intentional production with a domain in which there is the widespread ignorance that this is happening, and in which, according to the constructionist, there is widespread acceptance of an alternative, natural explanation of category differences.

---

[3] Because conscious intentional processes are resource intensive, drawing on finite resources like attention and willpower, this sort of shift from conscious and proximally intentional to unconscious and automatic is plausibly interpreted as a mechanism by which humans ontogenetically adapt to recurring environmental problems.

3 Revelation, Self-Knowledge, and Performance

The social constructionists I have been interested in here are not concerned with just any product of human intentional activity. Rather, the hope is for social constructionist explanations of phenomena for which there is a widespread belief or a temptation to believe that they are "natural"—that is, they are or were not under the control of human culture or choice. Such objects are covert constructions, and they contrast with overt constructions that are widely recognized as the products of human action—constructions like baseball games, marriage, money, and so forth. This revelatory aim of social construction is crucial for the social constructionist, but for performative accounts it produces a problem: the puzzle of intention and ignorance.

3.1 Ignorance as a failure of self-knowledge

In Chapter 4, I argued that a central reason that constructionists are interested in revelation hinges in part upon a plausible empirical assumption: that when a human kind is considered "natural" it tends to exculpate kind-typical behaviors. And some performative constructionists crucially hold that exculpating, natural explanations of behavior play a role in the intentional production of the behavior (e.g. Averill 1980a, 1980b) by changing the social contexts of rewards and punishments for the behavior.

But if kind-typical behavior is an intentional performance, how does this intentional performance go undetected? How is the illusion of naturalness maintained? We can elucidate this problem from both the third-person and first-person perspectives. From the third person, the problem is: how is it that a behavior that is intentionally performed in response to incentives goes undetected as such? From the first person, the problem is: if (many or most or all of) my category-typical behaviors are intentional performances, then why isn't this fact apparent to me?

Perhaps the most tempting approach to resolving this puzzle (and the one I pursue here) solves the problem from both perspectives at once by invoking an actor's failure of self-knowledge with regard to the mental states and processes producing her behaviors (Griffiths 1997). In the first-person case, if an actor is not aware that certain mental states caused her behaviors, she may sincerely disavow that they do. The agent's sincere disavowal of those mental states as causes of her behavior can then also undermine a third person's attribution of intention, by providing evidence against such attribution. Thus understood, performative constructionist accounts depend (in the first and third person) on a failure to understand the mental causes of one's own behaviors.

3.2 Alternatives to failures of self-knowledge

Why does appeal to a failure of self-knowledge of one's mental states seem so compelling? The problem is that no other explanation seems plausible. Consider two other strategies for how intentional behavior could be widely misexplained: the *malingering* model and *the collective bad faith* model.

As suggested in Section 1, the malingering model involves an agent intentionally enacting the features of a category in order to elicit social benefits, escape social censure, and the like. The malingerer accepts social benefits or avoids social costs by exploiting widely shared understandings about putatively natural categories. In contrast to a failure of self-knowledge, the malingering model suggests that there is a failure of third-person knowledge about the mental states of actors, but no failure of first-person access to the mental states. Actors simply acquire social benefits through deception.

Social constructionists are interested in categories for which revelation is possible, that is categories for which many (or most or all) category-typical behaviors, or many (or most or all) instances of a category are misunderstood as natural. Combining this with the malingering model, we'd get that many (or most or all) category-typical behaviors or putative instances of a category are the result of outright acts of deception.

Thus understood, the malingering model looks quite demanding upon actors since ongoing deception is quite demanding. Begin with the thought that in at least some cases of covert construction, actors are (on some developments of the performative constructionist model) enacting features of a role in order to acquire social benefits to which they would not otherwise be entitled or to escape social censure to which they would otherwise be subject. In such cases, third parties would be motivated to detect fraudulent cases to deny those benefits and permissions. Thus, actors may perform under critical scrutiny.

Richard Dawkins and John Krebs (1978) pointed out decades ago that opportunities for cooperation among our ancestors would give rise to selection for the capacity to deceive, and this, in turn, would create selective pressure for the capacity to detect signs of deception in others. And, in fact, real agents do employ a variety of clues to detect sincerity, including, for example, looking for signs that conscious control is being exerted (DePaulo et al. 2003), or monitoring the genuineness of emotional responses and the spontaneity of behavior (Verplaetse et al. 2007). Studies in experimental behavioral economics suggest that subjects have considerable success in detecting potential cooperative partners in contrast to those who would defect (e.g. Brosig 2002; Frank et al. 1993). And some evidence using videotapes from police interviews suggests observers can achieve accuracy in lie detection using such cues (Vrij and Mann 2005; Mann and Vrij 2006). All this suggests that successful outright lying, especially sustained over time, is somewhat difficult. Against this background of scrutiny, a strategy of deceiving others is potentially costly because of the difficulty of success, the possibility of lost benefits of (current and future) cooperation, as well as the possibility of punishment.[4] And this fact makes the malingering model implausible as a model of a common, widely instantiated category. In addition, this model is phenomenologically implausible as an explanation for common categories like gender or the emotions. It just does not seem, from the first person, as if we ordinarily enact kind-typical behaviors in order to create the impression of belonging to a category to which we do not really belong. That is, our ordinary, category-typical behavior is phenomenologically unlike, say, malingering (which could involve a quite specific intentional performance of symptoms).

So the malingering model of performative construction is problematic since widespread failure of third-person understanding of the intentional causes of behavior would be difficult to sustain, and also because the sort of self-knowledge allowed by the malingering model does not comport well with our experience of ourselves as enactors of kind-typical behaviors.

A different possibility is the collective bad faith model. This model suggests that where there is constructed performance, we already know that these thoughts or behaviors are not really natural. Applied to, say, gender roles, the idea might be that because we are all enmeshed in the system of rewards and punishments, prestige and ignominy that come along with gender, and because we perhaps have no very concrete ideas about the alternatives to this system, we simply perform and allow performance all the while recognizing that it is a kind of elaborate, everyday social theater in which we are engaged.[5] Because it abandons appeal to deception in the performance of kind-typical behaviors, it escapes the appeal to widespread, ongoing deception that characterizes the malingering model. In effect, the collective bad faith model abandons any failure of knowledge (self or other), replacing it with a widely recognized and seldom challenged social

---

[4] See Baumard et al. (2013) for a similar argument in the context of moral behavior, and von Hippel and Trivers (2011, section 2) for an extended argument to the effect that people are good at detecting deception.

[5] Actual tempting candidates for collective bad faith are very complex in ways that make unclear whether beliefs are instances of bad faith or of self-deception or motivated cognition. In a famous study of members of a doomsday cult, Festinger, Riecken, and Schachter (1956) suggest that members' beliefs exhibited the curious property of becoming stronger in response to challenges, an early illustration of cognitive dissonance.

order. Notice, though, that this model undermines the attraction of performative constructionist accounts since it effectively abandons the revelatory aim of social construction, implying that performative constructionists have nothing very general or important to teach us about ourselves (since we already know it). In addition, like the malingering model, collective bad faith remains implausible as a very common explanation on phenomenological grounds. In contrast with malingering, it ordinarily does not seem, from the first person, as if we act in order to create the impression of belonging to a category to which we do not really belong. In addition, we can add that it also does not seem, from the third person, that our exculpation of others is typically rooted in our complicity in a system of permissions and obligations (as it might be in the case of the exculpation of behaviors of members of overtly constructed categories—"he was just doing his job," we might say). These considerations suggest that the collective bad faith model is also implausible for common, apparently widely instantiated categories.

4 Failing to Understand Myself: Unconscious Causes and Failures to Locate

How could such a failure of self-knowledge occur? One possibility is that this failure stems from unconscious mental states that we fail to detect and identify. Below I argue that an appeal to unconscious mental states is implausible for performative constructionists. Instead, I suggest they should acknowledge that a failure of self-explanation results from a failure to locate one's mental state in an explanation of one's thought and behavior.

4.1 Unconscious mental states

Consider first the possibility that intentional but misexplained actions are the result of unconscious mental states. For instance, Robert Trivers (1976) has speculated that the evolution of capacities for deception and the detection of deception would also lead to the evolution of unconscious causes of behavior that might enable a deceiver to survive scrutiny:

> If... deceit is fundamental to animal communication, there must be strong selection to spot deception and this ought, in turn, to select for a degree of self-deception, rendering some facts and motives unconscious so as not to betray—by the subtle signs of self-knowledge—the deception being practiced. (vi)

Similarly, Paul Griffiths invokes unconscious mental states precisely as an answer to our problem: to explain the sincerity of actors in performative constructionist accounts of emotion.[6] Griffiths's way of doing so is by drawing on the productive analogy between the human mind and a computer found in contemporary cognitive science: when one thinks of the mind as computational, unconsciousness seems easy to model as simply the lack of access of one computer component or program to the information being used by another computer component or program. Griffiths writes:

> The computational unconscious can be used to give a relatively unproblematic account of the sincerity... The motivation and planning underlying the response is not accessible to introspection or verbal report. (1997, 153)

A similar solution seems implicit in some invocations of dissociation in syndromes like multiple personality. The DSM-5 says that dissociative disorders involve:

> A disruption of and/or discontinuity in the normal integration of consciousness, memory, identity, emotion, perception, body representation, motor control, and behavior.
> (American Psychiatric Association 2013, 291)

However we talk about such unconscious mental states, they suggest a particular way of understanding the failure of self-knowledge: the mental states causing the action are unconscious or inaccessible to the mechanisms involved in self-explanation.

I take it that these appeals to unconscious mental states are best interpreted as involving failures to detect and identify the relevant mental states (for example, to identify their functional role or content).[7] This conception of unconscious mental states figures in Trivers, and in the passages from Griffiths and the DSM-5 considered above.[8] Similarly, von Hippel and Trivers (2011) draw attention to implicit memories, implicit attitudes, and automatic processes—all of which look to involve mental states that are unconscious because of failures of detection and identification. We can see this idea elsewhere, as well, as in David Velleman's comment on the Freudian slip that "takes its agent by surprise because the motive is unconscious: he is not aware of wanting to do what he does" (2000, 12). And we could add to these other recent work on automatic processes that has focused on mental causes (e.g. associative processes or implicit biases) of

---

[6] Griffiths calls it the "problem of sincerity" that confronts what he calls "disclaimed actions."

[7] Conversely, see Robbins's (2006) discussion of successful introspective knowledge as involving detection and identification of one's psychological states.

[8] See Griffiths (1997, 151ff). Cf. Nichols and Stich (2003); Doris (2009). Because theorists often do not distinguish between detection, identification, and what I call location, it is not always clear what is intended. An admirable exception is Doris (2009), though he is clear that his primary concern is with detection (2009, 80, endnote 16). Though he doesn't draw this distinction Nahmias (2007) is, like me, clearly concerned about failures of location.

which we are not even aware (Doris 2009, 2015; Gendler 2008). In all these cases, unconscious states involve failures of detection and identification.

4.2 Two objections to unconscious mental states in performance

Unfortunately, an appeal to unconscious mental states by performative constructionists faces two serious objections. First, against a background in which we apparently routinely succeed in explaining our thoughts and behaviors, invocation of failures of access to the causes of intentionally performed behavior looks to be ad hoc. It explains how we could be sincere in self-explanations and overcome the scrutiny of others. But it doesn't explain why a particular case or sort of case should be a failure. What performative constructionists really need is some account of the specific circumstances under which mental states figuring in intentionally performed behavior are not consciously accessible—for example, an account of how a community's collective practice of employing a classification of a putatively natural category can result in failures to understand the mental causes of an intentional performance.

The second problem concerns whether agents in the performative constructionists' model can plausibly claim to be unconscious of the mental states that give rise to their behavior. Performative constructionist accounts posit an intentional response to widely held beliefs, beliefs that I suggested could be characterized as common knowledge. This fact—the fact that everyone knows them, and everyone knows that everyone knows them—makes it very odd to say that an agent is unconscious of having them.

Consider a simple belief-desire model. Then modeling production of multiple personality behaviors as intentional activity gives us a schema like Figure 5.1.

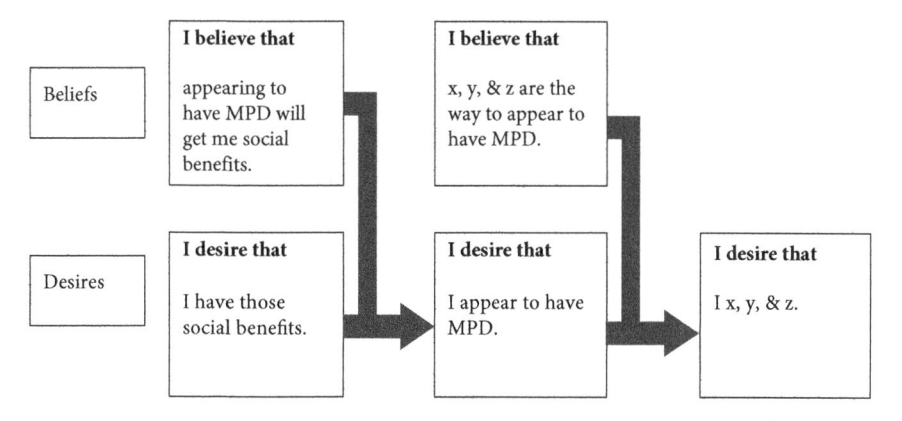

Beliefs	**I believe that** appearing to have MPD will get me social benefits.	**I believe that** x, y, & z are the way to appear to have MPD.	
Desires	**I desire that** I have those social benefits.	**I desire that** I appear to have MPD.	**I desire that** I x, y, & z.

Figure 5.1 Intentional performance of "multiple personality"

Now, consider the contents of the "belief boxes" in the top row. On the social role model, these beliefs are common knowledge, and so they are beliefs that everyone in the relevant community has. Because of this knowledge, the agent can perform symptoms leading others to identify the performer as a member of the category and react appropriately. Asserting that the agent does not have these beliefs is implausible since they are background knowledge of people in the community in which the performance takes place. Something similar might be said of the desires, namely, that it is obvious to everyone that the agent wants certain social benefits, ceteris paribus.

Together, these two objections—that the appeal to unconscious mental states is ad hoc and that it is implausible for social role performances—suggest that the performative constructionist should look elsewhere for an answer to the puzzle of intention and ignorance.

4.3 A failure to locate

Fortunately, unconscious mental states (understood as failures to detect or identify one's mental state) are not the only way to understand failures of self-knowledge in intentional but misexplained behavior. We can instead understand such failures as resulting from an inability to locate a mental state. A failure to locate a mental state is a failure to accurately represent its causal and rational role in our mental and behavioral economy. Failures to locate are compatible with successfully detecting and identifying the mental state, since one might know one has a mental state, and know what mental state it is, but not understand the causal and rational role it plays.

A psychological locus classicus for failures of self-knowledge resulting from both unconscious mental states and failures of location is Nisbett and Wilson's (1977) paper "Telling More than We Can Know." In it, Nisbett and Wilson review a range of psychological evidence of different sorts of failures of self-knowledge and self-explanation, but they also present an account of these failures that distinguishes failures of location from failures resulting from unconscious mental states.[9] That is, according to their explanation of some results, we sometimes detect and identify our own mental state, but fail to locate the state as the reason for and cause of our thought or action.

---

[9] This aspect of Nisbett and Wilson's account is substantively distinct from those implied by the contemporary dual process accounts that are its most obvious descendants, even though Nisbett and Wilson's work is typically presented as a seamless predecessor to contemporary dual process accounts (e.g. Haidt 2001), including by Wilson himself (Wilson 2002).

Consider Nisbett and Bellows's (1977) study that showed subjects unable to identify the factors that most strongly influence their own judgments of likability in an interview setting. In the study, subjects were asked to consider a candidate who had spilled coffee during the interview, and they judged the candidate to be more likeable. Subsequently, subjects didn't think the spilling of the coffee was part of the explanation for why they liked the candidate. Other results (e.g. Nolan et al. 2008) deliver the same verdict.

In such cases, there is a failure of self-knowledge. But it is not a failure that stems from failure to detect and identify the mental states that are causally responsible for the judgment. For example, subjects need not forget that the candidate spilled the coffee. Subjects just cannot introspect the causal role that that belief played in producing their judgment. On Nisbett and Wilson's view, these studies reveal a deep and important fact about the human mind: even where our introspective access to our mental states is intact, we have failures of self-knowledge because of our inability to access the processes connecting our mental states to one another and to our behaviors.

How can this apply to cases of performative construction? Looking back at Figure 5.1, the idea of failures to locate is that we, quite generally, fail to have introspective access to causal processes (represented by the arrows) that connect our mental states, even when we do have introspective access to the mental states themselves (the boxes). Subjects might know, for example, some general facts about multiple personality disorder. And they might even recognize their desire for certain incentives. What they don't have, on this model, is an understanding that those mental states are producing their own behavior.

Lacking such access, we instead construct a causal explanation of our thoughts and behaviors on the basis of our background theories about what is and is not a plausible cause. The result is that we might decide, along with others in our epistemic community, that our mental states and behaviors have completely different causes than they in fact do. Notice that if we fail to recognize a desire to be a product of instrumental rationality—if we consider our case like that in Figure 5.1 but without any knowledge of the causal processes (depicted as arrows) connecting the mental states—the mental states come to look quite different. If subjects don't realize that the desire to act out multiple personality disorder (MPD) symptoms, or, say, act out racial or gender or emotion templates, is instrumental to the satisfaction of other desires, then desires on which they instrumentally act may seem simply like primitive desires that subjects have as part of their natures.

4.4 Assessing the failure-to-locate model

The unconscious mental state explanation of failures of self-knowledge was both ad hoc and implausible, but the failure-to-locate model is not. It is not ad hoc because the performative constructionist has an explanation for why failures of self-explanation occur precisely in the cases of performative construction and not others. Nisbett and Wilson allow that we often succeed in explaining why we think or act in the way that we do. Their point is that such explanations are mediated by a posteriori causal theories that suggest to us what is a plausible cause of what. As with other sorts of explanation, our success at self-explanation is thus dependent upon the quality of our causal theory, and this will be substantially a function of our epistemic community. Where our theories are consonant with the actual factors that produce our judgments, we can accurately report on the factors that influence our judgments (Nisbett and Wilson 1977, 250ff; cf. Nahmias 2007). Conversely, where we have bad theories, our self-explanations are likely to fail.

Ex hypothesei in the cases that interest the performative social constructionist there is a widely accepted but false explanatory story about why we think and act the way that we do. So, for example, performative constructionists about gender might hold that gendered behavior is intentionally performed in response to widely held but false beliefs connecting biological category membership with those behaviors. But those same widely held beliefs are available to figure in a false explanation of those behaviors. When one behaves in a way that accords with the predictions of a false but accepted theory, one can be genuinely mistaken about why one acted that way.

And this account is compatible with the subject having access to the beliefs and desires that cause them to act in accord with the role—beliefs and desires that may be obvious to everyone in the community. According to the failure to locate model, they may both detect and identify the relevant mental states without thereby believing that those mental states explain their other thoughts or their behaviors.

4.5 Applying the model to the construction of gendered action

Even if this model applies well to an unusual category like "multiple personality disorder" there remains the question of how well the failure-to-locate model could work for an everyday category. Consider two gender cases to illustrate the model.

4.5.1 MALE PRIVILEGE AND MARITAL INFIDELITY

There is a small but persistent gap in marital infidelity rates between the sexes for which there are well-known evolutionary psychological explanations that emphasize different male and female reproductive strategies rooted in human biology (e.g. Shackelford and Buss 1996).[10] How could a performative social constructionist explain this gap using representations of a kind to explain kind-typical behavior? Suppose:

> *Michael the Philanderer*: Michael cheats on his spouse. When he explains himself, he is inclined to think and say things like "the heart wants what it wants." Or "that's the way we [men] are wired... It sucks but it's just a fact of nature. It's also tragic." Or "The problem with men is not whether they're nice or not. It's that it's hard for them at a certain point in their lives to stay true. It just is. It's almost not their fault. But it feels like it's their fault if you are involved with any of them."[11] That is, he explains his philandering as a product of a desire that he views as part of his masculine nature.

The performative constructionist model would explain Michael's philandering in part by reference to the role that beliefs about masculinity (including his own) have when they become common knowledge. Because Michael's views are widespread in his community, they tend to exculpate his behavior because of the plausible link between natural causes of behavior and mitigation of the reactive attitudes (see Chapter 4). Like everyone else, Michael knows this—he knows that, for example, he won't be judged so harshly for cheating as his wife would, and he knows that if he is caught, his wife will be judged negatively for failing to forgive him, for failing to accept this unfortunate consequence of his putative nature. Once we have this social context in place with a social role structured by an understanding that is common knowledge, the performative constructionist proceeds to explain Michael's behavior by appeal to the differential incentives created by the role (complementary reasoning could explain why Michael's wife is less likely to have an affair). Suppose this constructionist explanation is true.

These same representations of his behavior as natural also structure Michael's self-explanation. He is aware of the relevant contextual facts, but he does not believe that his recognition of these facts play a role in the production of his behavior. This allows Michael to sincerely explain his behavior by reference to his natural sexual desires while denying that his beliefs about unequal treatment play

---

[10] According to the General SS at the University of Chicago, marital infidelity rates over several decades suggest that the rate is about 12 per cent for men and 7 per cent for women every year (Castleman 2009).

[11] These explanations come from, in order, Woody Allen, explaining his relationship with Soon-Yi Previn in Isaacson (2001); the character Mike McConnell in the movie Adventureland by Mottola (2009); Nora Ephron on men and her ex-husbands, in Levy (2009, 68).

a role. He can know what he believes and desires without recognizing the role these beliefs and desires play as reasons for or causes of his behavior.

4.5.2 GENDER OPPRESSION AND FEMALE PASSIVITY

It is a fixture of feminist discourse that women may be socialized to accept subservience, which may in turn be used to rationalize a theory of women as naturally passive (e.g. Wollstonecraft 1995; Haslanger 2012). In recent decades, this theme has been developed in Catherine MacKinnon's work. MacKinnon offers an account of the role that patriarchal culture and the threat of violence against women play in sustaining male dominance (see e.g. MacKinnon 1987, 1989). While MacKinnon's view has been much discussed, and has very many nuances, one feature of it is something like a performative constructionist account of gender.

MacKinnon thinks that women's lives and behaviors are structured by social roles that are themselves enforced with the threat of violence. To be sure, MacKinnon thinks such coercion is far more ubiquitous than actual open threat of force (because, in our terms, the possibility of force is itself common knowledge). The result is that distinguishing consent from nonconsent becomes more difficult:

> The deeper problem is that women are socialized to passive receptivity; may have or perceive no alternative to acquiescence; may prefer it to the escalated risk of injury and the humiliation of a lost fight; submit to survive. (1989, 177)

If the threat of force is ubiquitous, operating even where no explicit threat of force exists, it offers women incentives to behave passively. Faced with a social role in which "passive receptivity" is the best available behavior, women may acquiesce. Crucially, however, on MacKinnon's view the desire for passivity also plays a role in self-explanation, and it offers something like an account of "false consciousness"—an account of why an oppressed person may claim to have freely chosen. Elsewhere she writes:

> Women who are compromised, cajoled, pressured, tricked, blackmailed, or outright forced into sex (or pornography) often respond to the unspeakable humiliation, coupled with the sense of having lost some irreplaceable integrity, by claiming that sexuality as their own. Faced with no alternatives, the strategy to acquire self-respect and pride is: I chose it. (1989, 150)

Consider how this fits the model of performative covert construction we have been sketching. According to MacKinnon, faced with a coercive situation structured by a conception of women as passive and of dominance as erotic, some women choose to acquiesce. This looks to fit well with the social role production of behaviors that I have been developing. But they come to understand their acquiescence not as the product of these background social factors of which they may be well aware (that is, not as the result of the threat of violence), but as the expression of their natural, primitive desires. If a woman acts as MacKinnon describes and says, "I chose it" of a behavior, it expresses the idea that "it is an expression of a desire that is primitively mine, not one that is instrumental to my satisfying some other desire such as the desire to avoid violence."[12] This is just how they would appear when subjects are unaware that the threat of violence is part of the reason for an instrumental desire to acquiesce.

5 Self-Explanation and Agency

Failures of self-knowledge—especially those that, as in MacKinnon's work, intersect with coercion—raise questions about agency, freedom, and responsibility. In this section, I argue that failures of self-knowledge undermine agency. Social constructionists are well known for being concerned with representations of putatively natural kinds of human, a tendency that leads them to be sometimes depicted as anti-intellectual and anti-scientific, particularly in evolutionary psychological and naturalist corners (e.g. Pinker 2002). Here, I again defend constructionist concern with representation by showing that false representations of the kinds to which we belong can undercut our agency in nonobvious ways.

5.1 Reflectivism, autonomy, and failures of location

Discussions of philosophically important properties such as autonomy, agency, freedom of the will, moral responsibility, personhood, and selfhood are shot through and through with a commitment to what John Doris calls "reflectivism": the assumption that people can (and even do) reflect and deliberate upon their mental states in deciding whether they constitute a reason for action (2009, 8ff; 2015). And as Doris notes, failures to detect and identify mental states that in fact influence your thought and behavior pose the quite general threat of undermining accounts of ourselves that depend upon our ability to reflect and alter our attitudes based on such reflection. For instance, systematic failures of detection and identification of our mental states leave in question whether we have the capacity to reflect upon, deliberate about, criticize, and modify our thoughts and

---

[12] MacKinnon's account suggests something more: that the selection of this explanation is motivated by the subject's preference that the explanation be true. While MacKinnon offers no evidence for her account, there is extensive literature on motivated cognition (Kunda 1999), and she may well be right. The present model is compatible with such motivated cognition, but silent as to its extent.

intentions that Charles Taylor (1976) has suggested are essential to our being responsible human beings.

There are substantial empirical questions about how widespread failures to detect and identify effective mental states are (Doris 2009, 2015).[13] Still, Doris is right about the pervasiveness of reflectivism in ethical philosophizing about agency (and related notions like autonomy, responsibility, freedom, humanity, and personhood and selfhood). It remains for us to add here that the required reflection is not only threatened by a failure to detect and identify mental states, but is also compromised by failures to locate them, and for largely the same reasons. This connection is registered by Eddy Nahmias, who offers a specific diagnosis of the problem with the failure to locate one's mental states as one's reasons: 'The less accurate your knowledge of which factors might influence you, the less you can control the influence of those factors you want to make a difference—that is, the less you can act on your principles' (2007, 10). In Nahmias's view, the problem with failing to know the causes of your actions is that it compromises your ability to make yourself into a person that can act in accord with your own deeply held principles—it undermines your capacity for autonomy.[14]

If Michael the Philanderer recognizes that his behavior rests on his recognition of the possibility of exploiting unjust social permissions, he might decide that acting upon such social permissions is indefensible because the permissions are rooted in falsehoods, or are unjust and unequal. Similarly, and this is surely part of MacKinnon's point in making these claims, a person who takes submission to be primitive rather than instrumental (i.e. instrumental to avoiding violence) loses the opportunity to consider whether the background reasons for submissive behaviors or their enforcement are ones that they endorse given their other values. Losing track of where one's beliefs or desires are "located" in one's scheme of reasons means that we lose the opportunity to undercut them by addressing them as background reasons, that we miss (as Nahmias suggests) a chance to make ourselves into the person that we want to be. Still, we must be careful not to assume that conscious agency is always good. If MacKinnon is right, then the success of a failure of location may itself be helpful in sincerely producing

---

[13] Doris is not optimistic about such identification, but in contrast to the present argument, this leads him to search for an antireflectivist theory (2015). Despite this difference, the present conception overlaps with his in that both are what he calls *collaborativist*: they envisage successful agency as emerging from social interaction.

[14] Cf. Washington and Kelly (forthcoming) for a similar externalist approach to the relationship of knowledge and agency.

behavior that enables someone to avoid violence. In the actual world, successful location may not always be a good thing for an actor, all things considered.

5.2 Covert construction and taking responsibility

We can see failures of location compromising agency in a different way if we consider its intersection with responsibility. Here I consider John Martin Fischer and Michael Ravizza's prominent compatibilist account of "taking responsibility" (1998). Strawson's idea that moral responsibility is integrally bound with our practice of the reactive attitudes has offered recent theorists of responsibility inspiration in offering a compatibilist account of moral responsibility. Fischer and Ravizza's development of the idea suggests that being responsible for an action requires in part that the mechanism that produces the action be "moderately reasons responsive" and that it be "an agent's own" (1998, 207). Specifically, Fischer and Ravizza require that an individual "must accept that he is a fair target of the reactive attitudes as a result of how he exercises this agency in certain contexts" (213) where that means,

> the agent takes responsibility for acting from a particular kind of mechanism... this theoretical characterization... makes more precise the intuitive idea that one takes responsibility for actions that spring from certain sources (and not from others). (215)

On Fischer and Ravizza's account, it is not that the individual must explicitly take responsibility for a particular act or mechanism in order for the agent to be responsible. Rather, "taking responsibility" comes in the implicit recognition on the part of an individual that products of a certain mechanism are "fair targets of the reactive attitudes" according to a given set of "social practices" (1998, 211, cf. 211–16). Taking responsibility for behaviors resulting from a particular mechanism or source amounts to having standing beliefs to the effect that behaviors from that source are appropriate targets of the reactive attitudes in one's community. Fischer and Ravizza offer this condition as a way of making sense of puzzles of action authorship, wherein if our intentional actions have the wrong sort of source (e.g. brainwashing), we rightly deny responsibility for them. But they go on to argue that such a conception is part of what it is to conceive of ourselves as one in a community of moral agents (1998, chapter 7).

Returning to performative constructionist accounts, on the model I have offered, it is the central and surprising claim of performative constructionists that some kind-typical behaviors may be proximally intentional, and therefore be what Fischer and Ravizza call "reasons responsive": if the agent's reasons were different, the agent would act differently in response to them. However, where an agent sincerely believes her action to be not the result of a reason's responsive mechanism, but instead a product of nature, a product of forces outside choice, the agent precisely fails to "take responsibility" for the action in the sense that Fischer and Ravizza emphasize. Indeed, in the case of a performative construction, an individual and the members of that individual's community share beliefs to the effect that behaviors from a particular source or mechanism are not appropriate targets of the reactive attitudes.

Thus, Fischer and Ravizza's account of taking responsibility, when paired with the possibility of covert construction, has a very surprising (though I think correct) result: that the capacity to act as a free agent with regard to actions of a certain type depends upon shared understanding of the sort of mechanism from which the action emerges. Widespread false accounts of a behavior-producing mechanism (of the sort that social constructionists suppose to obtain in a community) in fact genuinely undermine our capacity for responsible agency. Where we as a community think of a sort of action as "natural and inevitable"—where we collectively take the "objective attitude" towards an agent or type of agent with respect to certain thoughts or behavior—we in effect make it the case that we are correct in doing so, for the occupant of the social role that we thereby create produces behavior without "taking responsibility" for it. It follows that even empirical disputes over the proper representation of our own natures as humans and as members of various putatively natural categories can ramify in the shape of our agency.[15]

6 Constructing Explanations and Constructing Agents

I have articulated a model of performative constructionist claims that might be applied to specific categories, and explored its implications. My aim has been to make clear that such claims are both explanatorily and normatively relevant. They are explanatorily relevant in that they offer a distinctive and provocative account of human behavior situated in a social world that is structured by theories of what sorts of things we are, and they provide plausible models of at least some actual category-typical behaviors. They are normatively relevant in that, where such models are realized in everyday categories, they suggest that the accounts we collectively give of our own natures may influence the sorts of

---

[15] Fischer and Ravizza's account of "taking responsibility" has been criticized for implying subjectivism about the conditions of responsibility (Ginet 2006; Eshleman 2001; McKenna 2000). In contrast to the cases that concern these critics, the agents in constructionist examples have appropriately internalized the norms for what counts as a fair target of the reactive attitudes in their communities. One could worry that this is still insufficiently objective, but requiring more seems to require abandoning a Strawsonian approach to responsibility.

behaviors with regard to which we understand the causes, and with regard to which we are agents and responsible. If this latter argument is right, the way we represent ourselves can play a role in producing or undermining possibilities for agency and responsibility. The result is an account of how widespread practices of representing human categories results in groups for whom some possibilities are opened and others foreclosed without any widespread awareness that this is happening.

PART II

Realizing Social Construction

6

Social Construction and Reality

Social constructionist scholarship emerged from, and helps to support, an increasing appreciation of the contingent social and personal causes that determine our representations of the world, including those that emerge in our best sciences. While "social construction" labels a range of different positions, it is perhaps most persistently associated in the popular and academic imagination with the "science wars" and the endorsement of some sort of *global* antirealism. In contrast, the social constructionist explanations I have been developing in this book are *metaphysically moderate*: they are *local*, concerning only particular domains, and their obtaining within those domains is compatible with naturalism and with core doctrines of realism that they are often taken to threaten.

In this chapter, I turn toward this larger intellectual context, identifying and further articulating the perceived tensions between constructionist thought and what I call a *basic realist* picture of the relationship of our representations and the world. I then situate the metaphysically moderate social constructionist accounts that I have developed with respect to these tensions. My aim here is not to definitively answer each of these threats, a project that is engaged elsewhere,[1] but rather to show what metaphysically moderate constructionism assumes with regard to these threats, and what philosophical burdens it takes on. I conclude that metaphysically moderate constructionism is a kind of basic realism even with respect to those categories of which it obtains.

Here is how I proceed. In Section 1, I note three realist commitments—to the *literal* interpretation of discourse about a domain, to the *success* of that discourse, and to the *objectivity* of the facts in the domain—and how they relate to one another. In Section 2, I note the way in which social constructionist accounts of representations look to threaten the success of discourse in a domain, and consider what the realist must assume. In Section 3, I discuss the construction of categories (understood again as the metaphysical counterparts of our kind

---

[1] See, for example, Devitt 1991; Kitcher 1993; Hacking 1999; Kukla 2000; Boghossian 2006.

representations). Here, the apparent threat posed by social constructionist accounts of categories is difficult to make more precise, but it seems most obviously to concern the objectivity of the facts that we represent. But following Gideon Rosen (1994), I suggest that it is very difficult to articulate a notion of social construction that is both tempting to believe and not objective.[2] In Section 4, I offer an alternative account of the metaphysical threat posed by radical social constructionist claims: one that emerges not by denying what I have called basic realism, but from its attempt to treat mind-world determination as fundamental, and I note that, given the assumptions of metaphysically moderate construction, we ought to expect constructionist explanation to have special purchase for human categories. In Section 5, I consider how our understanding of basic realism is to be understood alongside the metaphysically moderate covert social construction of categories. I conclude in Section 6.

1 Basic Realism and the Constructionist Threat

Among philosophers, it is widely held to be consistent, and even desirable, to be realist in some domains, and irrealist regarding others. One can endorse, for instance, common-sense realism, scientific realism, intentional realism, moral realism, or realism about mathematical objects or modal facts, for instance. My interest is in endorsing a kind of realism about socially constructed human categories.

But realism and its denial are doctrines that take on different forms depending upon the domain under discussion, and the exact content of realism in a domain is often a subject of some debate. My specification of what I call *basic realism* is thus somewhat stipulative, but it aims to capture the realist doctrines of interest to us here.

As I understand it, basic realism involves a certain broad conception of the relationship between three familiar realist commitments regarding our discourse of a domain and the domain itself. I begin with the three commitments, and return to their relationship. The three realist commitments that I am concerned with here suggest that our discourse in a domain is *literal*, *successful*, and picks out facts that are *objective*.[3] The first requirement requires:

---

[2] Some years ago, Sally Haslanger brought Rosen's paper and its possible connection to social constructionism to my attention. See Haslanger (2012, chapter 6) for her own discussion of Rosen's view, social construction, realism, and objectivity.

[3] For presentations of realism incorporating versions of these, see Sayre-McCord 1988; Devitt 1991; Pettit 1991; Wright 1992; Rosen 1994; Mills 1998; Boghossian 2006; Miller 2014. The latter two

Literal: Statements in the domain are to be read literally (rather than as expressions, or metaphors, or useful fictions), and they are appropriately evaluated as true or false.

Denials of literalness result in forms of instrumentalism, conventionalism, and expressivism about discourse. Constructionist discourse does not typically reject literalness, but its denial is captured in the expression, common in the social sciences, that some variable is "just a construct."

A second realist commitment is:

Success: Assertions about a domain are true, or largely or approximately true, or at least increasingly approximate truth over time.

Success obviously admits of degrees of strength, and considerable wiggle room, but the central idea is clear enough. Consider, for instance, scientific realists (Boyd 1983; Kitcher 2002) or moral realists (Boyd 1988) or intentional realists (Fodor 1997) who are inclined to hold that our scientific or moral or intentional discourse is largely or approximately true, or is at least progressively approximating the truth.[4] Note that one implication of Success is that terms for a domain successfully refer (and co-refer over time) to kinds in the domain. One strategy for being an antirealist is, then, to deny successful reference (Churchland 1981; Stich 1983, 1996).

The covert category constructionist position we have been developing is in a special quandary with Success. On the one hand, the covert category constructionist wants to insist that socially constructed categories are real things that underlie the apparent causal and explanatory power of one or more important categories of contemporary life. But on the other hand, the covert constructionist wants to insist that some important existing (folk or scientific) theories of these categories are mistaken about their underlying nature. Insofar as realism for covert category constructions is possible, Success for the existing theory must be

---

roughly correspond to what are sometimes identified as an "existence" dimension and an "independence" dimension (see Devitt 1991; Miller 2014).

[4] Bas van Fraassen (1980, 1993, 1998) has influentially argued for constructive empiricism, the view that science aims only at empirical adequacy, where this is achieved "exactly if what it says about the observable things and events in this world, is true—exactly if it 'saves the phenomena'" (1980, 12). While van Fraassen joins the realist in taking the content of scientific theories to literally make reference to unobservable phenomena, he stops short of endorsing the unqualified statement of Success above. Van Fraassen's view suggests that scientists could (consistent with science) adopt a form of antirealism that denies Success with regard to those aspects of theory that make reference to unobservables. (Such an antirealist science is still compatible with a limited reading of Success on which truth, or approximate truth, or increasingly approximate truth, extends to observational claims.)

```
   theory         world
        ↖ epistemic  ↗
          constraint
```

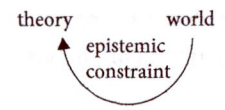

Figure 6.1 World-to-representation epistemic constraint

qualified: the category constructionist holds that we have Success in the midst of fundamental error. I return to this in Section 6.

The third commitment, *Objectivity*, is more poorly understood and more difficult to articulate. As a first pass, we can say:

> *Objectivity*: The world, or the part of the world that is in the domain of discourse, is mind-independent or "out there," independent of us.

What makes this requirement hard to specify is partly that it is usually agreed that *some sorts* of mind-to-world determination (or world-on-mind dependence) are metaphysically unmysterious and unproblematic. One challenge is, then, to divide these mundane dependencies from others. This challenge is especially pressing given social sciences including psychology, anthropology, and sociology where the objects of study look to be minds, or to essentially involve minds.

With these commitments in mind, we can note that realists assume a specific connection among these commitments, one we can depict in a simple way (see Figure 6.1). The arrow in Figure 6.1 indicates that, for the realist, when things are going epistemically well, the world causally and rationally determines the content of our theories of it. Thus conceived, the basic realist idea *combines* a commitment to literalness, success, and objectivity:

> *Basic realism*: The objective world in a domain causally and rationally determines or constrains the literal content of our theories of that domain, thereby explaining the success of those theories.[5]

Basic realism is not merely the conjunction of literalness, success, and objectivity for it embodies a conception of how these components of realism are related to one another.

Social constructionist work that came to be associated with the "science wars" seemed to threaten this basic realist conception in two different ways. The first was to suggest that, once we understand the social and historical processes involved in actual science, the causal and rational determination of our theories

---

[5] Moreover, the determination must be *of the right kind*. Here I ignore concerns about deviant determination not specific to constructionism.

by the world cannot be sustained, denying Success. Perhaps even more surprisingly, the second suggests that our theories or representations actually determine or "construct" the world itself; this determination reverses the supposed "direction of fit" of our representations with the world, threatening Objectivity.[6]

In the Introduction, I distinguished constructionist claims about *representations* from those about the *categories* that are represented. We can now add that this distinction that roughly corresponds to these two different sorts of constructionist threats to basic realism: social constructionist challenges to Success tend to focus upon the social determination of representations while social constructionist challenges to Objectivity tend to assert the determination of categories by our beliefs or collectively held theories.

2 Constructing Representations

Generally, when representational constructionists write of the construction of ideas, or theories, or understandings, or even more provocatively, of the construction of "knowledge" or "truth," they mean to pick out a causal process whereby some meaningful representation (rather than some other representation, or no representation at all) becomes an object of a belief-like attitude (rather than some other attitude or no attitude) by an individual or group.[7] Andrew Pickering's (1984) discussion of "constructing quarks," or Thomas Laqueur's (1990) suggestion that sex is "made up," or Foucault's (1978) claim that homosexuality first emerged in the nineteenth century are all concerned, at least partly, with the process by which *theories* of the quark or of sex or homosexuality are produced.

2.1 What constructs?

The causal processes that result in widely shared representations can take a wide range of forms. One telling dimension on which they differ concerns whether the causes are *impersonal* explanations that appeal to mechanistic causal forces or *personal* explanations that employ rational explanations.

Impersonal causes include cultures, conventions, institutions, or, more generally, circumstances that produce some representation. One influential version is Marx's claim that "The mode of production in material life determines the general character of the social, political, and spiritual processes of life. It is not the consciousness of men that determines their existence, but, on the contrary, their

---

[6] Boyd (1992) interprets social construction in much this way.
[7] Portions of this section are taken from Mallon 2013b.

social existence that determines their consciousness" (1904, 9–10). On Marx's conception, the material facts determine what people believe.

A different sort of impersonal determination comes in Thomas Kuhn's suggestion that, "what a man sees depends both upon what he looks at and also upon what his previous visual-conceptual experience has taught him to see" (1962/1970, 113), a suggestion with some foundation in "New Look" psychology (e.g. Bruner et al. 1951). A range of other authors subsequently took up this view. For example, the historian Thomas Laqueur writes that, "powerful prior notions of difference or sameness determine what one sees and reports about the body" (1990, 21). Provocative claims like Kuhn's and Laqueur's suggest the possibility that perception is so dependent upon background theories that the observational data becomes compromised as an independent constraint on empirical inquiry.

Other constructionist claims have emphasized *personal* explanations in which agents construct through their choices or actions. For example, Pickering's *Constructing Quarks* emphasizes the role of individual scientists' highly contingent judgments in a variety of roles in scientific process including, e.g., theory selection, experiment evaluation, assessments of research fecundity, and so forth. Such an emphasis on apparently highly contingent choices by researchers and scientific institutions is a mainstay of the social studies of knowledge.

Still other constructionists—those we might call *critical* constructionists—emphasize personal or group choices to highlight the role of individual or group interests or power relations in determining the content of an accepted representation. For example, Charles Mills discusses the possibility that the borders of American racial categories were determined in such a way as to

> establish and maintain the privileges of different groups. So, for example, the motivation for using the one-drop rule to determine black racial membership is to maintain the subordination of the products of "miscegenation." (1998, 48)

And a range of constructionist research, especially research on human classifications like "race" and "gender," documents shifts in human classification in response to shifts of interests or power.

While both impersonal and personal causes can serve to causally explain representational relations, they have somewhat different implications. Personal determination provokes questions about motivations and the responsibility for the representational relations, especially where the content of the representations may play an ideological role in harmful social processes. In contrast, impersonal determination can leave us seeming like mere continuants of impersonal causal forces, determined by what came before and determining what comes after.

2.2 Representational constructionism and scientific antirealism

A voluminous body of work in the history and philosophy of science shows that social, cultural, and idiosyncratic processes influence what contents come to be objects of belief or consensus, and detailed constructionist accounts of how certain representational relations come about teach us much about history and the practice of science (e.g. Latour and Woolgar 1986; Collins and Pinch 1993). But to say that the theory of quarks or that contemporary understandings of sex are "social constructions" is to say nothing about quarks or sexes themselves.

Such explanations come to seem philosophically provocative because they seem to undermine the realist idea that our theories are successful, or at least approximately successful, or are converging on the way the world is. Constructionist arguments undermining Success come in at least two forms: *incommensurability* arguments and *debunking* arguments.

2.2.1 INCOMMENSURABILITY ARGUMENTS

Acknowledgment of the messy empirical realities of the social determination of scientific belief has led to greater appreciation of the sometimes profound differences of outlook among proponents of competing or successive theories or paradigms. Rather than successive theories progressing toward greater and greater truth, incommensurability arguments suggest that science is, in Arnold Toynbee's phrase, simply "one damned thing after another."[8]

Methodological incommensurability arguments claim that there is no theory neutral way of adjudicating between competing or successive scientific theories, at least across substantial theory change, so that disputes over which theory better explains the evidence are inevitably question begging.[9] While methodological incommensurability arguments are powerful because they seem to depend simply upon the underdetermination of theory by data, they are also curiously out of touch with the general tendency of the scientific community to converge upon shared criteria of theory choice.[10]

In contrast, *semantic* incommensurability arguments suggest that theory change brings with it a change in the *meanings* of theoretical terms, so that new theories are literally about different entities than the old ones, rendering the idea of progressive accumulation of knowledge of a common subject matter as nonsense. Some of Kuhn's most famous and provocative language emerges from this understanding of the meanings of theoretical terms. For instance, he writes:

---

[8] Toynbee (1957, 268) coined this phrase to describe H. A. L. Fischer's account of history.
[9] See Kuhn (1962, 1977).   [10] See Laudan 1984, 87ff.

Consider... the men who called Copernicus mad because he proclaimed that the earth moved. They were not either just wrong or quite wrong. Part of what they meant by "earth" was fixed position. Their earth, at least, could not be moved. (1962/1970, 149)

Such semantic incommensurability was one reason why Kuhn made seemingly more radical pronouncements to the effect that: "the proponents of competing paradigms practice their trades in different worlds" (1962/1970, 150).

Nowadays, this semantic incommensurability has a well-worn rejoinder: co-reference over theory change can be ensured by adopting a different account of the reference of theoretical terms, one on which their reference is not mediated by the changing contents of the theories in which they figure, but is rather determined by some *external* or "causal-historical" relation to the referent which can be preserved through theory change.[11] As Richard Boyd, a pioneer of this new understanding, writes:

The anti-realist consequences which Kuhn (and Hanson) derived from descriptivist conceptions led to the articulation by realists of alternative theories of reference. Characteristically, these theories followed the lead of Kripke (1972)...and Putnam (1975)...Each of them advocated a "causal" theory of reference...It is by now pretty well accepted that some departure from analytic descriptivism, involving some causal elements, is a crucial component of a realist approach to scientific knowledge.   (2002, sect 4.1)

In light of this work, rapid changes in scientific theory look to be compatible with co-reference to the entities picked out by the theory, allowing a realist view on which successive versions of the theory more and more closely approximate the truth and the satisfaction of Success. In light of the fact of methodological convergence and the possibility of co-reference across theory change, constructionist arguments from incommensurability no longer seem a dire threat to Success (though covert category constructionists have special challenges in achieving co-reference, a topic I discuss in Chapter 8). Still, constructionist accounts of representation suggest another threat to the basic realist: causal exclusion of the world as a determinant of the content of our theories in a domain.

2.2.2 DEBUNKING AS CAUSAL EXCLUSION

Debunking constructionist arguments emphasize social and idiosyncratic determinants of accepted theories and the sometimes-arbitrary processes by which consensus is achieved. To the extent that these arguments successfully explain the content of our accepted theories, constructionist explanations threaten to

---

[11] Papineau (2010) defends realism without the assumption of co-reference.

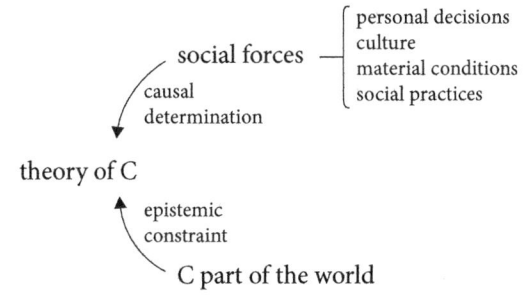

Figure 6.2 Social forces as an alternative explanation

```
                    social forces ─┬ personal decisions
                   ╱                │ culture
                  ╱ causal           │ material conditions
                 ╱  determination    └ social practices
                ↓
          theory of C
              ↑ epistemic
              ╲ constraint
               ╲
                C part of the world
```

causally exclude the basic realist account by appeal to an alternate explanation. After all, if idiosyncratic personal factors or social conditions explain why scientific consensus has the content that it does, there is nothing left over for the facts in the world to explain.[12] Constructionists thereby undermine our faith that our theories have the content that they do as a result of causal, rational determination by the way the world is.

Thus, some constructionist work seems directed at undermining the rationality or veridicality of the scientific processes by revealing the arbitrary, or perhaps ideological, character of the causal processes by which some researchers become leaders, some research programs come to be widely pursued, and some representations come to be widely accepted. Instead of broadly rational inquiry that leads scientific representations to increasingly accurately reflect features of the natural world, constructionist work can seem to suggest that our theories are determined by processes that we have little reason to believe are reliable indicators of the truth (see Figure 6.2).

Notably, the so-called "strong program" of the sociology of scientific knowledge (sometimes called "the Edinburgh School") was at the center of much of the "science wars" controversy because of the threat of such exclusion. David Bloor (1976) argues against the tendency to use sociological explanatory strategies *only* in the explanation of *false* theories, or of *peculiar* features of the history of science like "Kepler's mystical beliefs about the majesty of the sun" (7). Bloor instead emphasizes the importance of what he calls *symmetry* in sociological explanations: if social, environmental, or idiosyncratic causes operate to produce *false* beliefs, they must also figure in the explanation of *true* or *successful* beliefs as well: "the same types of cause would explain, say, true and false beliefs" (Bloor 1976, 5).

---

[12] The terminology of "causal exclusion" is adopted from Kim's work on mental causation. See, e.g., Kim 1998.

Similarly, Thomas Berger and Peter Luckmann, who popularized the term "social construction" with their book *The Social Construction of Reality*, write that, "the task of the sociology of knowledge is not to be the debunking or uncovering of socially produced *distortions*, but the systematic study of the social conditions of truth" (1966, 12; italics mine). To the extent that social constructionists offer explanations of our belief-like representations on which they have the content that they do (rather than some other) due to causes that bear no obvious connection with the facts our beliefs purport to represent, constructionist explanations threaten to debunk those beliefs. Answering this sort of argument in particular cases looks to require justification of deviations from symmetry that show one is not simply engaged in the ad-hoc rationalization of widely accepted theories.

2.3 Realism and multi-factorial determination

The debate over the implications of the social determination of scientific theory has now played out over decades and, as a result, nowadays a great deal of attention is paid to the actual history and practice of science. Philosophers of science now widely acknowledge a central causal role for social and historical events and processes in the determination of the content of scientific decisions of all sorts, and realist philosophers of science typically believe that these social processes can be integrated into more sophisticated accounts of the way that the scientific community produces widely accepted, successful theories, and can even figure as part of the explanation of that success (e.g. Boyd 1992; Kitcher 1993; Longino 1990).

While I focus no further on debunking arguments here, the success of discourse about a domain assumes that the partial determination of representations by "social processes" (and other debunking determinants) is at least sometimes consistent with those representations also being partially rationally and causally determined (in the right way) by facts in the part of the world that they represent. This assumption is widely shared among both realists and empiricists in the philosophy of science and is not especially controversial.[13] For reasons that have been part of philosophical discussion since Plato, *global* endorsement of debunking of beliefs seems to be self-undermining. If social determination of belief (or widespread belief) is always debunking of the content, and all such belief is so determined, then this belief too would be debunked. More *local* debunking remains possible, but local debunking requires careful study and analysis to assess the

---

[13] For instance, for a constructive empiricist like van Fraassen (1980) the rational and causal constraint is imposed by observation rather than by the world, but it is constraint all the same.

relative contributions of various causal determinants (Kelly 2011; Nichols 2014). Exactly this sort of analysis can (and often does) vindicate nondebunking determinants of our representational attitudes, making qualifications to the symmetry of the sociological explanation of representations not ad hoc.

3 Category Construction

Category construction of the sort I have focused on in much of this book poses a very different challenge to basic realism. Here the worry is that features of our minds, social practices, history, or collectively held theories are in some metaphysically troubling way determining features of the world. Following work in philosophy of science (e.g. Boyd 1992), I'll call this vague assemblage of views "neo-Kantian constructionism."

Recall (from the Introduction and Chapter 2) that human category constructionists hold that human mental states, decisions, culture, or social practices that represent a category *C* causally or constitutively explain how *C* came to have instances, or how *C* continues to have instances, or how members of a category came to have or continue to have their category-typical properties. Because the control of a category by our representations of it seems to call into question the objectivity of the category, reversing the presumed "direction of fit" of our beliefs, concepts, or theories with the world, such claims demand some account of the construction of categories that makes clear the process, or the mechanisms, or perhaps the "metaphysical picture" on which representations of categories cause or constitute categories.

3.1 What does objectivity require?

While social constructionist claims can seem to suggest that the world (or an aspect of it) is not objective, it is hard to say exactly what objectivity requires. There *is* wide agreement that realism is committed to something like it. Michael Dummett writes:

Realism I characterize as the belief that statements of the disputed class possess an *objective truth-value, independently* of our means of knowing it: they are true or false in virtue of *a reality existing independently of us.*   (1963/1978, 146, italics mine)

And Hilary Putnam similarly suggests:

What the metaphysical realist holds is that we can think and talk about things as they are, *independently of our minds*, and that we can do this by virtue of a "correspondence" relation between the terms in our language and some sorts of *mind-independent* entities.
(1982, 141, italics mine)

And Ernest Sosa (1993) claims:

What the metaphysical realist is committed to holding is that there is an in-itself reality *independent* of our minds and even of our existence, and that we can talk about such reality and its constituents by virtue of correspondence relations between our language (and/or our minds), on the one hand, and *things-in-themselves* and their intrinsic properties (including their relations), on the other. (609, italics mine)

Alexander Miller (2014) characterizes a position he calls "Generic Realism":

a, *b*, and *c* and so on exist, and the fact that they exist and have properties such as *F-ness*, *G-ness*, and *H-ness* is (apart from mundane empirical dependencies of the sort sometimes encountered in everyday life) independent of anyone's beliefs, linguistic practices, conceptual schemes, and so on.

While realism is committed to independence of the domain in question, it is also often agreed that this requirement of realism is consistent with *some* mind-world determination so long as those are limited to what Miller calls "mundane empirical dependencies." Such allowances are generally meant to permit very specific sets of entities to count as unproblematic. Michael Devitt, for instance, writes that "The world that the Realist is primarily interested in defending is independent of us except in one uninteresting respect. Tools and social entities are dependent on us" (1991, 249). And C. S. Jenkins remarks that: "mundane kinds of dependence on the mental are to be ignored when realism is characterized as mind-independence." She then adds, "We should, however, acknowledge that it is not straightforward to say exactly which kinds of dependence *are* mundane" (2005, 199).

Supposing that this is right, and that at least some human-dependent causal or constitutive facts are consistent with satisfaction of the sort of objectivity required by realism. It follows that if we can understand seemingly radical claims of social determination as involving only such "mundane empirical dependencies," then these would not run afoul of realism by being inconsistent with the required need for objectivity. On such an understanding, category constructionism involves mundane causal and constitutive claims in need of further explanation, but these claims do not provoke any special methodological or metaphysical handwringing, for they do not threaten the basic realist commitment to an objective world.

3.2 Distinguishing moderate and radical construction

Nonetheless, there remains the venerable thought that some sorts of mind-to-world determination are of a radical, problematic sort. Transcendental idealists are the paradigms of this sort of determination, and many have interpreted the social determination of belief or collective belief as generating more general worries about the social constitution of the world, thereby undermining requisite objectivity.

Many initially, and perhaps unfairly, interpreted Kuhn in this way though Kuhn himself seems to have grown more "neo-Kantian" and less naturalist in later years (Kuhn et al. 2000; Bird 2002). The later Kuhn writes:

> By now it may be clear that the position I'm developing is a sort of post-Darwinian Kantianism. Like the Kantian categories, the lexicon supplies preconditions of possible experience. But lexical categories, unlike their Kantian forebears, can and do change, both with time and with the passage from one community to another... Underlying all these processes of differentiation and change, there must, of course, be something permanent, fixed, and stable. But like Kant's *Ding an sich*, it is ineffable, undescribable, undiscussible. Located outside of space and time, this Kantian source of stability is the whole from which have been fabricated both creatures and their niches, both the "internal" and the "external" worlds. (2000, 104)

Similarly, Judith Butler (1990), Nelson Goodman (1978), Hilary Putnam (1990), and Richard Rorty (1979, 1998) all say things that invite some version of a neo-Kantian interpretation upon which our concepts, descriptions, and social practices invariably constitute or partially constitute the facts that we study.[14] In light of this influential tradition, it would be useful to have a clearer idea of what demarcates the border of these metaphysical badlands, to be sure we haven't wandered over.

Consider some mundane cases. I turn the key to start my car. We form a Neighborhood Watch group to reduce crime. I pay my taxes on time to avoid penalty. All of these cases presumably involve mind-to-world determinations of a completely unproblematic sort. By comparison, some social constructionists make claims regarding the construction of scientific facts, or sex, or race, or emotion. What makes *some* constructionist claims provocative while others are mundane?

3.3 Radical construction as noncausal determination?

One tempting thought is that the realist allows for mind-to-world determination that is causal, but not that which is noncausal. The *causal* determination of some sorts of facts of the world by minds is widely regarded as unproblematic. For instance, Gideon Rosen writes:

> Let us say that an item depends *causally* on the mind iff it is caused to exist or sustained in existence in part by some collection of everyday empirical mental events or states. Mind-dependence of this sort is perfectly familiar. And no one supposes that the observation

---

[14] For critical discussion, see Kukla 2000; Bird 2002; Alcoff 2006; Boghossian 2006; Haslanger 2012.

that an item is causally mind-dependent in any way undermines its claim to objectivity in the sense that interests us...If the Kantian imagery is to be made sense of, it must therefore be possible at the very least to sketch a species of mind-dependence distinct from this sort of causal dependence. (1994, 287–8)

Similarly, Andy Egan (2006) writes:

Certainly a fact's being *causally* mind-dependent—in that its obtaining was brought about by some mental activity—does not make the fact metaphysically second class. Facts about the existence of artifacts, for example, are as metaphysically respectable as facts get, and they are causally mind-dependent. (102–3)

Since the realist clearly wants to allow for mundane causal influences of mind on world perhaps we can (as Rosen suggests) characterize the constructionist as insisting on the complementary idea that the mind noncausally determines the world. This is the strategy Boyd pursues in characterizing the challenge posed by constructionism in the philosophy of science. On his characterization, the scientific realist asserts and the radical social constructionist denies: "*The No Noncausal Contribution Thesis*: Human social practices make no noncausal contribution to the causal structures of the phenomena scientists study" (1992, 172). This strategy seems promising since it both plausibly interprets constructionist claims in a way that looks to render them continuous with Kantian philosophical influences while at the same time offering an explanation of the mysteriousness that such views seem to carry with them: they involve some sort of noncausal determination.

One problem with using the concept of causation to demarcate the difference between problematic and acceptable mind-to-world influences is that it is not clear that accounts of causation are up to the task of neatly dividing mundane, *moderate* determination from *radical* determination of the world by our mental states or theories. Treatments of the metaphysics of causation are too numerous to mention, and far too numerous to consider one by one here. But we can at least get a sense of the difficulties.

Perhaps the most widespread family of accounts of causation are counterfactual accounts that understand C causing E as E's counterfactually depending upon C (Lewis 1973), or perhaps counterfactually co-varying with C (Lewis 2000). Prima facie, counterfactual dependence lets in too much to rule out radical construction. Consider some worlds:

White Noise World has minds and other stuff, but the other stuff has no "joints." The properties of the stuff are evenly distributed throughout the stuff. Categories in this world are merely "imposed" upon them by the minds. What it is to be a chair or mountain in this world is simply to be a group of sensory experiences of the stuff upon which a classification has been imposed.

Berkeley World has only minds with sensory experiences. What it is to be a "chair" or a "mountain" in this world is simply to be experiences with the proper organization.

Kant World has both a real in itself (noumenal) aspect and a mental (phenomenal) aspect. But all the ordinary facts of our experience are constituted as part of the phenomenal aspect, products of our minds "imposing order" upon the noumenal world.

Transigent is a world in which most facts obtain or don't obtain depending on whether they are thought to do so. If a mind (or in *Transigent** a group of minds, or in *Transigent*** a mind or group of minds) of the right type thinks "There is a chair" or "There is a mountain," then there is a chair or a mountain.

Even if each of these worlds is possible, they seem in some way radical, violating the realist picture on which the facts in the world are appropriately independent of the minds that think about them. Nonetheless, in each world the existence of chairs or mountains is counterfactually dependent upon mental states. By hypothesis, in each case, we could alter the chair and mountain facts of the world by altering the mental states. It follows that counterfactual dependence need not rule out radical sorts of mind-to-world determination.

Consider, in contrast, a more restrictive account of causation. For instance, some theorists have attempted to analyze causation in terms of a "transfer" of energy or some other physical quantity (e.g. Fair 1979; Salmon 1994). This sort of account *does* seem to rule out radical sorts of determination presumably underlying neo-Kantian determination, and those in the worlds above, insofar as we have no sense of what sort of "transfer" could be in operation. But the cost of drawing our border with metaphysical badlands here is to rule out too much. It is far from clear that all the presumably "mundane" relationships that we want to count as causal would be so under such a transference account.

Reflect on overt conventional facts. On David Lewis's (1969) account of conventions, conventional facts depend upon an elaborate arrangement of dispositions to think and act in certain ways, conditional on the preferences, dispositions, and knowledge that one takes others to have. It is not clear how to understand the relationships between these dispositions and conventions in terms of, say, energy transfer. Or consider artifactual facts, and assume that artifacts are in part constituting as artifacts by the intentions of those who make or use them. Again, it seems implausible to understand the relationship between the intentional states that constitute an object as an artifact and the object so constituted as an instance of such transfer.[15] So long as we regard a failure to regard conventional facts or artifacts as real to be a mistake, we need some other, less strict account of causation.

While there is much more that could be said here, and said more carefully, I now turn elsewhere. I offer as an interim conclusion only that it is unclear that appealing to causation can distinguish radical from moderate determination of the world by the mind.

3.4 Rosen, quietism, and the Martian anthropologist

The difficulties with finding a way of distinguishing mundane empirical dependencies from radical determination leads some to the temptation to think that there is something confused about connecting realism with objectivity at all. Gideon Rosen (1994) has pressed this point by appeal to what I call *human naturalism*—the view that humans are features of the natural order, and that our properties belong among the objective properties of nature.[16] Rosen argues that once we accept a naturalist picture of the human mind, features of these minds including their relations (for example, any dispositions that they might have to cause or to "project" properties onto the world) are themselves perfectly objective features of the universe that belong in any "absolute conception of the universe." Rosen's argument begins with the observation that:

Most of the good philosophers writing in the 19th century took it for granted that *the world as a whole* was in some sense psychic—penetrated through with thought or mentality—and hence that the study of Mind was the proper foundation for the study of absolutely everything. These days, of course, we can hardly take the idea seriously...
The Mind of the idealists was, after all, a very peculiar thing by our lights: an entity not quite identical with anything we encounter in the natural world—and this includes the "subject" of empirical psychology—which nonetheless somehow constitutes or conditions that world. And the trouble with idealism is that we just can't bring ourselves to believe in this Mind anymore. A flexible and relatively undemanding naturalism functions for us as an unofficial axiom of philosophical common sense. This naturalism is so vague and inchoate that any simple formulation will sound either empty or false. But it is a real constraint: and one of its

---

[15] These objections resemble more general worries about transference accounts. See Laurie Paul and Ned Hall (2013, section 3.3.5) for discussion.

[16] There are, to be sure, real and unresolved questions about how some of our properties, especially our mental ones, can be located in the natural order. And some of these properties—the intentional ones—I make use of in what follows. Nonetheless, I bracket worries about naturalizing the mental. If intentional properties and activities are ultimately resolved in favor of a thoroughgoing and unproblematic naturalism, then they can be taken for granted. If they ultimately prove to be in some way unassimilable to the natural world, then the argument will show that social construction has no metaphysical problems above and beyond those attendant on naturalism about intentionality itself.

implications is that if we believe in minds at all, they are the embodied minds of human beings and other animals. Our most basic assumptions thus leave no room for the transempirical Subject whose relation to nature was the urgent problem of post-Kantian metaphysics. And since it is just plain obvious that empirical, embodied minds do not actively constitute the bulk of animate nature, the idea that the world as a whole is in some sense mental can only strike us as an incredible fantasy. (1994, 277–8)

Rosen argues that such human naturalism has no serious philosophical alternatives in contemporary philosophy. It's not that we cannot describe alternatives; rather, it is that we do not and cannot believe them. He goes on to argue that against the background of this contemporary naturalism, it is not only nineteenth-century idealism that loses its grip on us. On his view, the twentieth-century realism/antirealism debate, insofar as it requires an alternative to objectivity, also ceases to make any sense.

As we have suggested, realists imagine the world as *really there, objectively* and not because of our thinking about it or inquiring into it. Some neo-Kantian antirealists seem to deny this, holding that the world (or at least certain parts of it) is a "projection," or is "subjective," or "dependent on our minds." As Rosen puts it, "We can epitomize the realist's basic commitment by saying that for the realist as against his opponents, *the target discourse describes a domain of genuine, objective fact*" (1994, 278, italics in original). And he later writes, "the objectivity that interests us is thus evidently to be contrasted with a sort of mind-dependence—or to use a more modern idiom, a dependence on our linguistic and social 'practices'" (287).

This idea of dependence can be manifest in different ways. Kant has it that the world we can have knowledge of is in part constituted by the conditions on the possibility of that knowledge. Social constructionists like Kuhn sometimes gesture at just such conditions stemming from the social determination of knowledge, or from the ineliminability of socially contingent language, theories, and concepts. Extending Rosen's arguments to these descendants of Kant requires only small and plausible steps. Suppose, for example, that we follow Rosen in considering the relevant kind of mind-dependence or theory-dependence involved in category construction as a sort of *response-dependence*. A schema for response-dependent properties might be of this sort:

x is a member of category P iff a creature c with an M-type mind would classify x as P in ideal epistemic conditions.

We can easily extend further the dependence to languages or cultures:

x is a member of category P iff a creature c with an M-type mind, with an L-type language, and a J-type culture would classify x as P in ideal epistemic conditions.

Rosen's point is that whether or not some x satisfies such a schema is as objective a fact as any other. Therefore, the truth of such dependence schemas does nothing to show a metaphysical contrast with other claims. There is nothing "irreducibly subjective" about them. Once we extend Rosen's argument in this way, a focus on Objectivity seems to fail to provide a metaphysically deep distinction between the real and the socially constructed.

One way Rosen captures his point is with the idea of an anthropologist from outside the community whose mental states or social practices "construct" some fact—what I'll call a *Martian anthropologist*. If an anthropologist from Mars were to come to Earth and fail to document the many mind-dependent, culture-dependent, and language-dependent facts that characterize human life, the Martian anthropologist would have missed out on perfectly real facts of human life on Earth. With this image in mind, it is not clear that there is a way of stating a kind of mind dependence that we ought to regard as nonobjective in the relevant sense.

Rosen casts his own view as a kind of quietism (e.g. 1994, 279; cf. Miller 2014), but it might rather be rendered as a consideration in favor of realism about entities that exhibit various forms of human dependence. Rosen's argument seems to show not that Objectivity turns out to make no sense, but rather that the conditions for objectivity are satisfied in any of the worlds we take to be plausibly ours. It suggests that constructed facts are as objective as more paradigmatically natural facts. If constructed facts run afoul of realism, it is not by running afoul of Objectivity.

4 Moderate Construction

What is at issue, then, between metaphysically radical constructionists and metaphysically moderate ones? Rosen's discussion suggests it could be simply an atavistic conception of the place of mind in the universe. But while that may underlie some of our concerns about Objectivity, it does not seem to exhaust our concerns. As Miller (2014) points out, more than human naturalism is needed:

> Suppose we found out that facts about the distribution of gases on the moons of Jupiter supervened directly on facts about our minds. Would the threat we then felt to the objectivity of facts about the distribution of gases on the moons of Jupiter be at all assuaged by the reflection that facts about the mental might themselves be susceptible to realistic treatment? (Miller 2014)

Miller's point is that such a scenario is troubling even if we accept human naturalism.

But we could also add that the threat from Miller's scenario has little to do with the objectivity of the relevant facts. Rather, it seems troubling simply because it fails to be explicable in terms of more basic, or better understood mechanisms or causal principles. Given Miller's scenario, if we subsequently found such a mediating mechanism by which our minds determined the distribution of gases, the resulting dependency might be a curiosity of science, but not a metaphysical mystery.

I propose that this is the remaining core division between radical social constructionists and metaphysically moderate ones. Radical social constructionists are willing to take mind-to-world (or social-to-world) determination to be an unexplicated metaphysical primitive.[17] Metaphysically moderate social constructionists are not so willing.[18] In the remainder of this section, I further contrast metaphysically moderate constructionism with radical constructionism, and then with metaphysically moderate nonconstructionism.

4.1 For metaphysical moderation

Radical constructionists about categories differ from moderates simply regarding the character of the *causes*, or alternatively, the character of that which constitutes, or provides the supervenience base of, or metaphysical grounds for, socially constructed facts.[19] On this construal, the metaphysically moderate constructionist insists, against the radical constructionist, that such causes or grounding facts are not metaphysically fundamental. For the metaphysically moderate constructionist the "mundane empirical dependencies" involving entities like mental states, representations, social practices, and so forth are "mundane" and "everyday" because we take these entities to be situated in, and nonfundamental aspects of, the broader natural world. For the moderate, constructed facts depend in familiar ways upon entities, mechanisms, processes, and relations that are themselves familiar, and these may in turn depend upon other, more basic

---

[17] In a similar vein, Devitt contrasts the "explicability" of mundane creations with the "ineffable" processes that constructionists depend upon.

[18] Metaphysical fundamentalism of this sort has a venerable history in defining realism itself. Michael Dummett, for instance, writes: "A dispute over a realism [regarding a given subject matter] may be expressed by asking whether or not there really exists entities of a particular type—universals, or material objects: or, again, it may be asked, not whether they exist, but *whether they are among the ultimate constituents of reality*. From this second formulation, it is apparent that opposition to realism often takes the form of a species of reductionism: certain entities are not among the ultimate constituents of reality if they can be 'reduced' to entities of other types" (1963/1978, 145, italics mine).

[19] Put in grounding terms, the central issue is, as Jonathan Schaffer (2009) has put it, not about what is real but about "what grounds what." Aaron Griffith (ms) urges understanding social constructionist claims as grounding claims.

entities, mechanisms, processes, and relations of familiar kinds (though surely we often lack a good understanding of the details of such dependencies).

In contrast to this presumed dependence, the radical constructionist holds, as Boyd puts it, that: "in some deep sense the structures studied by scientists are imposed on the world, in the sense of being reflections of the conceptual schemes they employ" (1992, 132), and he later writes that "the whole idea is that certain social practices *impose*, in something like a logical or conceptual way, a certain general causal structure on the world" (188). This imposition is, for the radical constructionist, the primitive process by which we make up the world that we come to know. Thus conceived, the radical constructionist insists that, among the basic and irreducible facts of the world are facts connecting mental states or representational contents or social practices to apparently nonmental entities.

If this diagnosis is correct, then metaphysically moderate constructionism is to be preferred to its radical cousin on the grounds of the metaphysical simplicity of its ontology and the explanatory power of its theories. Modern science offers a startlingly powerful theoretical apparatus in which there is a cascade of more and more basic explanations, suggesting more and more powerful and exceptionless universals, going right down to the laws of physics. The metaphysically moderate constructionist understands social determination in terms of entities and events that determine facts in naturalistically ordinary ways while the metaphysically radical constructionist suggests that we "add on" laws or other forms of determination, connecting minds or social groups or practices to nonmental facts in the world. As against the remarkable power and relative simplicity of the theory and ontology of science, these special mind-world relationships look to be instances of what J. J. C. Smart (1959) called "nomological danglers"—peculiar "add-ons" to the causal structure of the world revealed by modern science.[20,21] That is good reason to dismiss them unless there is explanatory work that only they can perform.

4.2 Moderate construction versus natural determination

Hacking (1999) has emphasized the special character of human kinds, arguing that they are interactive kinds because they exhibit "looping effects": a process by which mind-to-world determination and world-to-mind determination produce a causal loop that can lead the causal relata to evolve over time. This has, in turn,

[20] Smart attributes the term to Herbert Feigl (1959).
[21] Notice that the *global* radical constructionist who holds that *all* facts are (in some way) socially constructed achieves comparative simplicity in theory and ontology (despite taking on other, unattractive consequences) since the fundamental sort of determination in such a world is mind to world.

given rise to discussion of how to make the distinction between interactive and noninteractive (what he calls "indifferent") kinds more precise, and of whether all or only human kinds are interactive kinds.[22] Without attempting to settle these questions, we can note that constructionist claims regarding human kinds have a prima facie plausibility that similar claims about many other features of the world lack precisely because human beings are cultural animals for whom language and other sorts of symbolic expression provide central ways of cooperating and coordinating with others. Human traits, dispositions, and behavior are sensitive to the decisions, representations, and social processes that figure in constructionist explanations. Because of this, the social construction of kinds of human, action, and social arrangements has enormous plausibility, in contrast to the construction of mountains or quarks whose responsiveness to constructive causes is less clear. This plausibility warrants a special focus on the social construction of human kinds (as in this book).

5 Realism about Social Construction

Insofar as a category constructionist can sustain the Success of a theory about constructed entities, and offer a plausible account of their construction, the category constructionist can satisfy a basic realism about those entities. But this "realist" labeling needs clarification and qualification. It needs clarification because it seems to many philosophers to be a terminological mistake: realism and constructionism are by definition not compatible. And it needs qualification because the category constructionist needs to understand basic realism in a specific way. Consider these in turn.

5.1 Construction and the language of "the real"

Many philosophers consider the contrast of "constructivism" or "constructionism" with realism to be a fixed point, perhaps even a matter of definition. In much of contemporary metaethics, for example, construction is contrasted with realism precisely in virtue of the sorts of kinds or mechanisms that constitute, or provide the supervenience base, or the grounds, for properties in the domain. For many, if *moral wrongness*, for instance, is grounded in culturally local conventional facts, then this would be a kind of constructivism and not a kind of realism.

One might think that this is a deep dispute over what properties are entitled to be labeled with the language of "the real," or "reality." I do not think this is right. For instance, ethicists need not, and often do not, deny that ethical constructivists

---

[22] E.g. Bogen 1988; Ereshefsky 2004; Cooper 2004; Tsou 2007.

treat constructed moral properties as real or that mental states, culture, language, or social practices are real things. Russ Shafer-Landau writes that "constructivists endorse the reality of a domain, but explain this by invoking a *constructive function* out of which the reality is created" (2003, 14). Similarly, in race theory, Mills distinguishes racial constructivism from racial realism but he nonetheless remarks that "race may not be real in the sense that racial realists think or would even like, but it is real enough in other senses" (1998, 66).

In contrast to this trend of defining an opposition between realism and construction, Geoffrey Sayre-McCord has suggested that in ethics, those who hold that moral claims depend upon subjective or intersubjective facts can also be realists (1988, 15–16). I suggest the same holds for socially constructed human categories. The divide between facts grounded in human decisions or mental states or practices and facts grounded in other features of the universe is not a metaphysically deep one. Humans and their minds, cultures, decisions, social practices, languages, institutions, and arrangements are simply natural mechanisms among others that can subserve and sustain real categories.

5.2 *Covert category construction and basic realism*

While the considerations we have been posing suggest that there is no conflict between construction and basic realism, in this book we have been especially focused upon a particular sort of social construction: covert category constructions. And covert category constructions seem to run afoul of basic realism in two ways.

What the covert category constructionist thinks is that widely shared but mistaken theories of these categories influence (by causing or constituting) the world, giving rise to some of the very phenomena those theories represent. As we have already mentioned, this implies that where there is covert construction, Success needs to be qualified (for success occurs in the midst of fundamental error about the character of the kinds). But it also implies that basic realism incompletely depicts what success folk theories enjoy. Basic realism holds that the success of our theories is explained by their appropriate constraint by the world. But in cases of covert construction, the qualified success of the mistaken theory is also explained by the (metaphysically moderate) control of parts of the world by our representations. And this seems to violate the idea that the facts the (mistaken) theory represents are "objective" or "out there."

The important thing is to distinguish the existing, *mistaken theory* that the covert category constructionist holds figures in the construction of the world from the *constructionists' theory*. In a world in which the covert constructionist is correct, the widespread mistaken theory experiences qualified explanatory success but involves unacknowledged influence upon the represented world by its representation. Thus, with regard to the mistaken theory, the constructionist is partially irrealist (for the constructionist qualifies the success of the mistaken theory). Still, the constructionists' theory can explain the qualified success of the mistaken theory by appeal both to the reality of what it represents and to its role in causing or constituting the world. Thus, the constructionists' theory is compatible with basic realism without qualification. It does not need to qualify success, for the constructionist is (ex hypothesei) correct about the social mechanisms that structure the relevant human categories in the world. And it does not need to qualify objectivity, for the constructionist acknowledges and represents the influence of the mistaken theory upon the world it represents; such influence is central to the constructionist's explanation of the category.

In this respect, the constructionist is like the Martian anthropologist. The Martian's representations of the facts of humanity, we imagine, are relatively causally isolated from the domain of knowledge. (This is the point of imagining the Martian as a Martian, as *outside* the constructing group.) She observes from afar, listing the objective, mind-dependent facts of human life, but without interfering. For the Martian, too, the basic realist picture is correct: the fact that a group of *humans*' representations exert influence that is unrecognized by those humans is among the many objective facts that the Martian successfully registers.

At this point, many with constructionist sentiments will recognize an assumption that they are skeptical of: the assumption of a "view from nowhere," a position from which the thinker or scientist observes and theorizes without causally interacting with the object of her study.[23] We can allow this view is an idealization, but still retain it for some purposes. We do so first, because this idealization is an approximately realized regulative ideal for many actual cases, and second, because this idealization is already intrinsic to the covert constructionist conception of the social world.[24]

In noting that the idealization is approximately realized, we note that in many cases social scientists or theorists *have little or no effect* on their objects of study. It is a standard feature of methodology and critique in the human sciences to worry about the effects of experimenter representations upon the subject population, and thus we have methodological rules and protocols designed to address phenomena like experimenter demands or iatrogenic symptoms. As naturalists, we take such manipulations to be important precisely because they successfully

---

[23] E.g. Harding 2015.
[24] Rosen (1994, 284) similarly endorses the regulative role of such an ideal of objectivity.

approximate the realist conception that we can study the world "out there," as it is when we are not looking at it.[25]

In any case, conceiving of the constructionists' theory as of a piece with the Martian anthropologists' also fits with the constructionists' own conception of covert construction. As we have conceived it, the covert constructionist has a theory (the constructionists' theory) that relates a community, a mistaken theory, and the category or categories that the mistaken theory helps to structure. So it's part of the covert constructionists' theory that the community of study has the *mistaken* theory and not the constructionists' correct theory. In that respect, the constructionists' relevant causal isolation from the community she studies is part of the theory of covert construction. If the constructionists' conception of the mechanisms underlying a human category were widely adopted, it might undermine those mechanisms.

6 Covert Category Constructionism as a Sort of Basic Realist Theory

Metaphysically moderate constructionism has the aim of insisting that some things exist or have the character that they do because of mechanisms that are part of, or are under the control of, human minds: mental states, cultural information, human actions, and artifacts. Properly understood, this locates them squarely as competitors for the explanation of human categories like race, sex, gender, mental illness, the emotions, and so forth, and, as such, they figure as part of the so-called "human nature wars" rather than the "science wars." They oppose not basic realism as we have been understanding it, but instead particular accounts of these categories that explain human difference, especially cognitive and behavioral tendencies and dispositions, by appeal to natural (perhaps especially biological, neurological, or genetic) facts. And they do so by offering alternative causal and constitutive explanations.

But to say that the metaphysically moderate construction of human categories has prima facie plausibility is not to say that it is actually true. Prima facie plausibility only goes so far, especially in the face of biological, medical, nativist, and evolutionary accounts asserting alternative causal mechanisms. Making

---

[25] This may not put the point of the causal independence of constructionist theorizing strongly enough. I find it easy to believe that an academic could write books and articles about constructionism that few read and fewer believe and that, in any case, have no systematic, widespread effect on the categories in question.

social constructionist claims more plausible and more precise requires going beyond general thoughts about metaphysical plausibility to more and more specific accounts of the construction of categories, a task I have only begun in this book. It also requires answering other, general worries about the character of social reality that threaten to undermine the basic realist conception of success. It is to this task that I now turn.

7

Achieving Stability

> Heraclitus is supposed to say that all things are in motion and nothing at rest; he compares them to the stream of a river, and says that you cannot go into the same water twice.
>
> (Plato, *Cratylus*, 402a)

> It is evident that the state is a creation of nature, and that man is by nature a political animal. And he who by nature and not by mere accident is without a state, is either a bad man or above humanity.
>
> (Aristotle, *Politics*, 1253a)

If covertly constructed human categories are to support attempts at knowledge, prediction, explanation, and intervention, then it must be possible for them to be stable enough to undergird folk or social scientific knowledge of the social world. But a significant strain of social theorizing suggests otherwise. According to this strain, culture is so malleable that culturally produced kinds cannot be the objects of knowledge in the way that natural kinds can. This worry takes a number of forms. One worry is that human choices constitute an open-ended causal system that is susceptible to many forms of interference and may exhibit a sensitive dependence on initial conditions.[1] Or that, in particular, because of conceptual innovation and conceptual diversity, types structured by human intentional states will be highly volatile and local.[2] In one statement of such a view, the philosopher Charles Taylor writes:

> The success of prediction in the natural sciences is bound up with the fact that all states of the system, past and future, can be described in the same range of concepts...This conceptual unity is vitiated in the sciences of man by the fact of conceptual innovation, which in turn alters human reality. (1971, 209)

This worry about instability is only exacerbated when we look to constructionist social theory itself. Constructionist social theorists routinely assert the instability

---

[1] E.g. Taylor 1971.   [2] E.g. Fay 1983; Taylor 1971.

of social categories. For example, in their *Racial Formation in the United States*, Michael Omi and Howard Winant argue that race is "*an unstable and 'decentered' complex of social meanings constantly being transformed by political struggle.*"[3] Omi and Winant are not speaking here only of the *meanings* or *concepts* of race but also of the social roles that are structured by such meanings or concepts. And, like Taylor, they argue that the reason such social roles are unstable is that they are not grounded in biological facts, but rather are created via a "social and historical process."[4]

A more recent source of such pessimism is Hacking's thesis of "the looping effect of human kinds." Hacking draws attention to the possibility, important in the account of Chapters 2 and 3, that persons classified in a certain way may come to change in response to the labels placed upon them. According to Hacking, this "looping effect" destabilizes human kinds, and our knowledge of them, and he argues that looping effects mark "a cardinal difference between the traditional natural and social sciences" because "the targets of the natural sciences are stationary" while "the targets of the social sciences are on the move" (1999, 108).

If these philosophers and social theorists are right, then constructed human categories may not have the stability to ground our inductive, predictive, and practical success in the social world. How can we have successful theory regarding a constantly changing subject matter? How could we converge on the truth about it over time? Insofar as we take our explanatory and predictive success in understanding the social world to involve such stable ground, some other alternative—perhaps that provided by biological human categories—would have to be found.

There is surely something right about the emphasis that Taylor, Hacking, and many other social theorists place on the instability of human kinds. Indeed, the emphasis on such instability is part of the attraction of constructionist theorizing, for constructionist theorizing recognizes the cultural and historical contingency of phenomena that theorizing focused on biobehavioral kinds is thought to miss. What we need is an argument that allows for the variability and flexibility of culture, but also allows that constructed human categories can achieve sufficient stability to (at least sometimes) figure as the metaphysical subject matter of successful epistemic projects. Below, I explore the character of the problem that instability poses, and I suggest a number of mechanisms that can figure in securing the stability of constructed human categories, including mechanisms rooted in the very looping effects that Hacking discusses.

---

[3] Omi and Winant 1986, 68, italics in original. Cf. 1994, 54–5; 2014, 110; both without italics.
[4] Omi and Winant 1994, 55; 2014, 110.

I begin, in Section 1, by offering some very general considerations regarding Taylor's concerns with instability, and about the possibility of knowledge of a changing world. Then, in Section 2, I develop Hacking's thesis of the "looping effect of human kinds." In Section 3, I argue, using causal pathways that we identified in Chapter 3, that there is considerable reason to think that in many cases the psychological and social mechanisms appealed to by constructionists will stabilize, rather than destabilize kinds. In Section 4, I offer some evolutionary speculations about why this should be so. In Section 5, I conclude that the instability of constructed kinds need not pose a threat to the success of our theories of the social world.

1 Converging on the Truth or Wandering, in Search of the Truth?

The familiar, basic realist picture given holds that there is a mind-independent world "out there," and our successive theories provide an increasingly accurate account of it (Chapter 6). Within this broad picture, we can locate various kinds of realism. Crucially, in imagining progressive theoretical success, we imagine that the world *holds still*, while our theories, and especially our scientific theories, become more and more closely coupled to it. If the *entrenched social role account* of a human category is correct, the truth is that the category is caused or constituted by our representations of it. But if these social theorists are correct, then social reality, unlike those aspects of reality described by the hard sciences, does not "hold still." As such, it is unclear whether the realist conception of a theory that (perhaps increasingly) approximates the truth can obtain.

1.1 The Heraclitean social world

In Taylor's picture, we still have the realist idea of a mind-independent world that exists and constrains theory, but the social part of that world is controlled or constituted by human intentions and actions. On his view, intentions and actions are themselves constituted by concepts and beliefs that frequently change. This undermines the convergent coupling of our theory to the stable world that is possible for the natural sciences. Where the world is evolving on its own, it comprises a moving target for theorizing (see Figure 7.1).

On Taylor's picture, even if a theory were a *perfectly accurate* representation of the social world at a particular time (as theory$_2$ is, suppose, for the social facts$_1$ at t_1), change in the social world results in the theory becoming inaccurate (as theory$_2$ could be with respect to the social facts$_2$ at t_2). For this reason, Taylor

ACHIEVING STABILITY 165

Figure 7.1 Social change produces Taylor Instability

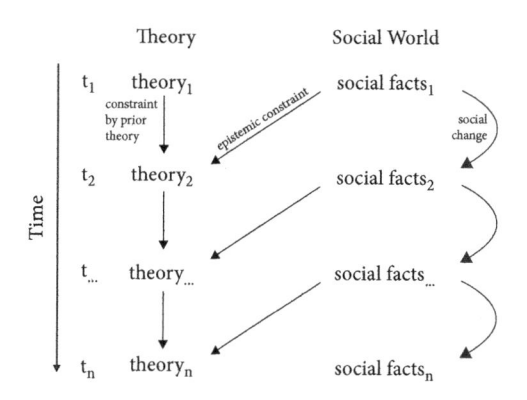

[Figure 7.1: A diagram with "Time" axis pointing down on the left. Two columns labeled "Theory" and "Social World". Theory column shows t_1 theory$_1$ (constraint by prior theory), t_2 theory$_2$, $t_{...}$ theory$_{...}$, t_n theory$_n$, with downward arrows. Social World column shows social facts$_1$, social facts$_2$, social facts$_{...}$, social facts$_n$, with curved "social change" arrows on the right. Diagonal "epistemic constraint" arrows go from social facts to theories.]

thinks, the facts we find in the social world do not support our epistemic projects in the way that the apparently eternal kinds of physics or chemistry do. Call this view:

> *Taylor Instability*: Concepts and conceptions (at least partially) constitute social reality at a time, but concepts and conceptions change frequently. Social reality changes whenever those concepts and conceptions do, undermining knowledge.

1.2 The possibility of knowledge of the changing world

The fact that the social world changes does not, by itself, mean that our scientific theories cannot become more and more accurate accounts of it. Our scientific theories might be less like arrows, aimed at a target that slips out of the way, and more like guided missiles, "homing in" on the moving targets at which they are aimed. The threat that Taylor Instability poses to our ability to rely on scientific theories as accurate accounts of the world depends upon the relation between the *rate* at which our scientific theories grow increasingly accurate and the rate of change of the social world.

The right-to-left arrows in Figure 7.1 indicate appropriate "epistemic constraint"—the rational, causal determination of our theories by the world in their domain. The left-side top-to-bottom arrows represent the constraint imposed on the content of our new theories by our belief in the content of our old theories. The top-to-bottom arrows on the right indicate Taylor Instability in the social world: changes that, ceteris paribus, tend to decrease the accuracy of theories that may have been true of some previous time slice of the social world in question. But if increases in the accuracy of our theories happen at a greater rate than decreases in the accuracy of our theories produced by changes in the social world, then overall convergence of our theory on the "moving target" of the social world will still happen. If they occur more slowly, it will not.

Consider knowledge of evolving natural kinds like species or varietal kinds. Such kinds, understood as instances of property-cluster kinds, exhibit typical properties. For instance, bird guides offer idealized descriptions and depictions of the typical features of various species, sexes, and varieties of birds, like this one of the mighty Northern Cardinal:

Cardinalis cardinalis
8 ¾ in. [22 cm]. *Male*: An *all-red* bird with pointed *crest* and black patch at base of heavy, *triangular reddish bill*. *Female*: Buff brown, with some red on wings and tail. *Crest, dark face*, and *heavy reddish orange bill* distinctive. *Juvenile*: Similar to female, but with blackish bill. VOICE: Song is clear, slurred whistles, repeated. Several variations: *what-cheer cheer cheer*, etc. *whoit whoit whoit* or *birdy birdy birdy*, etc. usually two-art. Call a short, sharp *tik*... HABITAT: Woodland edges, thickets, towns, gardens, feeders.

(Peterson 2010, 312)

The members of these species kinds possess kind-typical traits, resembling one another in certain respects in part because they share common ancestors in a process in which genes and other resources are inherited by new members of a biological population (or group of populations) that is characterized by barriers to gene flow with other biological populations (or groups of populations). Over time, selection pressures upon these populations change, genes come into and go out of the gene pool, and other resources may shift. This has the effect of changing the properties in the clusters that are typical features of the kind. In millions of years, descendants of the Northern Cardinal might have profoundly different typical properties.

Species kinds (understood as instances of homeostatic property clusters) change their typical features over time, and even come into, and go out of, existence. (Or: they come to have, and cease to have, any instances.) Naturalists can have knowledge of members of these changing kinds that allows us to engage in successful induction, prediction, explanation, and intervention because our capacity to gain accurate knowledge of these kinds can (sometimes) be far more rapid than the processes that underwrite biological change.

This rate of increasing accuracy is itself a contingent fact, and one that itself has changed over time. We can imagine a species of scientists that converge upon true theories of the world at a rate much more slowly than the changes in species and varietal kinds that occur over evolutionary time. To such beings, the biological world might make little sense, or the patterns they discern might relate only to broader phylogenetic patterns that seem epiphenomenal to us. Alternatively, the biological world might, to them, begin to seem like the biological world seems to a scientist studying a fast-reproducing system—a scientist like the biologist Richard Lenski whose populations of E. coli have (as of this writing) gone through over 64,000 generations since he began keeping them in 1988.[5]

This is just to illustrate that mere change or evolution in the world does not undermine knowledge of it, so long as the rate of change is less than the rate at which our theories improve themselves. Generalizations involving social role kinds and biological kinds will never have the broad domain that laws governing basic physical kinds do. But such limitations do not rule out knowledge of social role kinds any more than they rule out knowledge of other changing kinds like biological species. The test of whether a particular social role amounts to an important kind is whether making reference to it is useful in epistemic projects like explanation, prediction, and intervention. If what you are trying to explain or predict or intervene upon is itself local to a historical or spatial social setting, then generalizations that operate only in that narrow domain are unproblematic. These considerations are very general, but they offer us some background rules for understanding how claims of instability bear upon claims of the instability of knowledge of human kinds.

1.3 Alternative conceptions of kindhood

The problem of stability might be thought to be easily addressed with a shift in how we individuate constructed human categories. If we think of human categories as individuated not by the homeostatic property cluster that supports induction about the category, but exclusively by relational properties that play no such role, then we could allow the causally relevant features of category members to transform arbitrarily while preserving reference to a stable kind. In the case of the Northern Cardinal, for instance, we could say that though the properties of the typical Northern Cardinal change over time, the ancestors and descendants of the Northern Cardinal are nonetheless members of the continuing category.

This can be accomplished in a number of ways, understanding category members as:

- A member of a biological population that is characterized by reproductive isolation from other populations.
- A part of a spatiotemporally extended individual that consists in members of the category.
- A member of a lineage, connected by relations of reproduction.

---

[5] An overview of the E. coli Long-term Experimental Evolution Project Site can be found here: <http://myxo.css.msu.edu/ecoli/index.html>.

These alternative ideas are closely related to one another, and versions of each have been defended as accounts of biological species kinds.[6]

Crucially, on each of these accounts, whether one is a member of the population, or a part of the individual, or a link in the lineage does not depend upon the inductively relevant properties one has being the same or similar to fellow category members. That is, it is theoretically possible that two individuals could be members of a population or parts of an individual or members of a lineage and not be similar enough to one another to support induction.

In such accounts, the category becomes a bit like an individual in a dream or in some change-blindness experiment whose features are completely transformed though we continue to regard it as the same thing. On each of these accounts, the typical, causally significant properties of members of a kind at a time could change to an arbitrarily large degree without undermining the existence of instances of the kind or our ability to refer to it.[7] Moreover, with regard to biological species, there is now a rich literature regarding different ways of conceiving of species kinds. This includes accounts similar to the above construals as well as accounts advocating some sort of pluralism. I have little doubt that similar views can be articulated for human kinds.

But none of them are the right kind of solution for the problems I have been posing here. The instability we are worried about is not only instability in the *existence* of the category under transformation of its features, but rather an instability in the typical properties themselves, knowledge of which allows reference to the category to be useful in our explanatory and inductive projects that concern such (temporally and spatially bounded) entities. In contrast, treating human category terms as picking out exclusively relationally characterized populations, or spatio-temporally extended individuals, or historical kinds characterized by a lineage of reproduction offers no assurance that reference to the kinds will support inferences over even short amounts of time. Because our problem throughout this book has been understanding the social construction of kinds in a way that allows for them to be causally powerful, our worry about their instability is a worry about understanding how culturally produced or sustained kinds could play that role. A conception of a kind that allows us to continue to

---

[6] The first is Mayr's classic species concept (1984). Michael Ghiselin (1974) and David Hull (1976; 1978) have suggested that the second is true for species. Ruth Millikan has offered an account of historical kinds that suggests (3) (1999). See Bach (2012) for an application of this idea to gender. Michael Hardimon (2003) asserts that the assumption of something like (3) is part of the logical core of race concepts.

[7] A qualification is in order since there's some overlap between the typical, causally significant properties of a kind and the properties that underlay biological reproduction. I ignore this complication here.

refer to a kind in the absence of any ability to support induction, prediction, and explanation is not a solution to that worry.

Acknowledging this point still allows for two possibilities. First, it may be the case that a pluralistic approach to human kinds is appropriate: we can and ought to pick out human kinds differently for different purposes. I think this is the case, but I do not argue for it here. Second, it is also possible that we could *combine* historical conditions and property clusters in our account of which features of a kind are necessary to it.[8] Whether we wish to do this, whether we wish to regard homeostatic property-cluster kinds as nonhistorical types or as partially constituted by spatio-temporally located particulars, will depend upon features of our specific explanatory interests in the context in which we are using terms, features from which I abstract here. What we cannot do is simply abandon commitment to a more robust co-instantiation of properties on pain of losing the inductive power of human categories.

2 The Social World and Looping Effects

So genuine knowledge of social kinds seems possible, but is it actual? What we need is some further indication as to whether the sorts of social role kinds that we have been sketching can be stable enough to play a role in our inductive projects, or are, rather, so unstable relative to our capacity to accurately pose hypotheses and develop theories about them that they undermine the possibility of knowledge. Before we consider that question further, I turn to consider Hacking's work on the "looping effect of human kinds," and his distinct concerns regarding the instability of social kinds. As we saw at the outset of this chapter, Hacking emphasizes both the capacity of our conceptions of human kinds to constitute a social world that we represent, but also of the inherent instability of such a social world. He emphasizes both when he writes: "People classified in a certain way tend to conform to or grow into the ways that they are described; but they also evolve in their own ways, so that the classifications and descriptions have to be constantly revised."[9] Here, Hacking suggests that the social world evolves partially in *response* to our epistemic activities with regard to it. And while he recognizes that this change can lead to greater conformity, he suggests that in the main it leads to the need for constant revision. Writing of child abuse, he suggests

---

[8] Some theorists suggest a mixed view on which homeostatic property-cluster kinds are characterized partially by historical properties (e.g. Millikan 1999; Griffiths 1999; Wilson et al. 2007). Relatedly, Devitt (2008, 2010) understands species kinds as explained partially by the (he believes intrinsic) properties that explain them and partly (in some cases) also by historic properties.
[9] Hacking 1995a, 21.

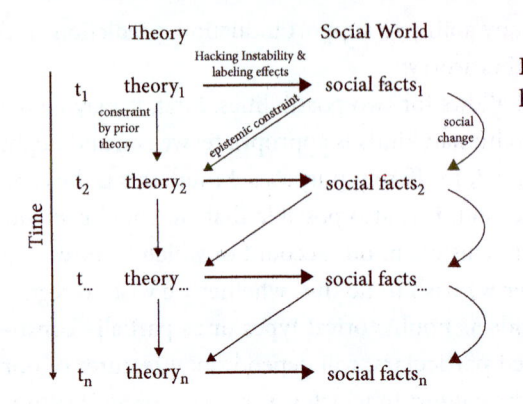

Figure 7.2 Hacking instability and labeling effects

that "The concept of child abuse may... be so made and molded by attempts at knowledge and intervention, and social reaction to these studies, that there is no stable object, child abuse, to have knowledge about."[10] If Hacking is correct, the influence of our theoretical activities on the social world involves a complicated and conflicted set of forces (see Figure 7.2).

Like Taylor, Hacking emphasizes that some sorts of instability are simply products of the role that concepts play in structuring the social world. Once a conception of a kind structures a category, changes in the conception may disrupt the category it structures. For instance, Hacking suggests that the theoretical linking of fugue to hysteria in late nineteenth-century France, and the subsequent skepticism about hysteria as a medical category in the early twentieth century, helped to undermine the social conditions that made individuals undertake fugue behavior (1998, 71ff). This is not a case where the world changed, and so our theories became untrue. It is rather a case where the theories themselves changed, causing the regularities that those theories produced in the world to cease to be produced.

What Hacking's picture adds to Taylor's is explicit reference to a causal pathway—from our theories, understandings, and other representations of the world to the social world they describe—that produces instability. Hacking emphasizes that changes in the social world that are produced by categorized persons reacting to the theory that a social group applies to them.

In light of our discussion in Chapter 3, we note that this putative causal pathway plausibly includes multiple causal mechanisms. Importantly (as Hacking himself recognizes), it includes *both* forces that lead categorized persons to

[10] Hacking 1995a, 61.

conform to and therefore *confirm* the theory that describes them, but also forces that lead them to *deviate* from the theory and therefore *disconfirm* the theory. Hacking suggests that the latter forces, together with the sorts of changes Taylor emphasizes, result in targets of social science that "are on the move." Hacking seems to allow that human kinds may be relatively stable across instances at a particular time, but suggests that over time human kinds will be inevitably destabilized by our epistemic and practical projects. We can call this:

> *Hacking Instability*: Theories of human categories lead the humans who putatively fall into such categories to react in ways that disconfirm those theories.

As we observed in Section 1, whether or not we have knowledge of a category that is "on the move" depends, in part, on the rate of our knowledge acquisition with regard to it, and the rate of change in the typical properties of category members. Hacking adds to this picture the idea that our projects of knowledge acquisition are themselves a causal factor that may produce change in (and potentially affect the rate of change of) our targets of knowledge.

2.1 Modeling stability and instability

Suppose some new theory of a human kind or a set of human kinds comes about: race, or child abuse, or homosexuality.[11] One thing that can happen is that this new theory can lead people to act *in accord* with the new theory. Following Thomas Scheff, we can call these *labeling effects* (Scheff 1974). Labeling effects influence social reality, but they do not bring it about that theory becomes falsified by reality. Rather, the two become more tightly coupled. As such, this form of categorical instability does not cause our theories to become undermined by changing social reality. It does not threaten the success of our theories.

Alternatively, a new theory could lead people putatively in the labeled class to act *at odds* with the new theory. One way this could happen is that collective action or individual opposition by members of a labeled category could take the form of exhibiting alternative sorts of behaviors (acting against a dominant conception of the category), undermining the accuracy of the conception of the labeled category. Michael Walzer describes that in the caste system of ancient

---

[11] Before abandoning the term "human kinds," Hacking used it to mean "kinds of people, their behaviour, their condition, kinds of action, kinds of temperament or tendency, kinds of emotion, and kinds of experience" (Hacking 1995a, 351–2; 2007). Elsewhere (Mallon 2003), I have argued that we ought to interpret his claims about the instability of kinds as limited to covert social role kinds. Hacking's own work focuses upon covertly produced kinds, and these kinds are the most likely to vindicate Hacking's claims of instability.

India, "A certain kind of collective mobility is possible, for castes or sub-castes can cultivate the outward marks of purity and (within severe limits) raise their position in the social scale" (1983, 27). Other examples can be found in the myriad microsocial interchanges that comprise contemporary social life, as when, for instance, Brent Staples, an African American man walking around Hyde Park in Chicago, describes his practice of "whistling Vivaldi" (a deviation from expectations) to disrupt the fright that some white passersby feel towards him (Staples 1986). Such acting against category conceptions, writ large, does destabilize categories in ways that threaten the representations of those categories with falsification and obsolescence.

Hacking, however, does not just seem to have acting against category in mind, so much as the possibility that a new theory of a human kind may alter the social conditions that give rise to the dispositions, traits, or behaviors that it attempts to describe. For instance, Hacking speculates that as new theories of child abuse become disseminated, they create a new social context in which agents may choose different actions, including new and different forms of abuse (1986; 1995a, chapter 4, 1999, chapter 5) Abuse statistics may then rise or shift as new people (and sorts of people) decide to abuse, or as new forms of abuse take hold.

A third kind of reaction Hacking emphasizes is that putative category members might organize in response to a social role and attempt to alter it by pressuring those who produce and broadcast the theories that structure it. While homosexuality was once regarded as a disease, those classified as homosexuals organized and systematically agitated to alter the norms (including the practices of discriminatory treatment) that go with the label (e.g. Hacking 1986). Hacking documents similar phenomena occurring with multiple personality sufferers (1995a), and those classified with the label (and their relatives) of childhood autism (1995b, 1999). However, as debunking representational constructionists emphasize, we should be careful here. A conception can change not for epistemic reasons, but because of political pressure—for instance, pressure upon those who count as experts about a category. When this happens, it is not a case where changes in the world result in our conception becoming false. It is rather a case where other social forces lead to changes in our theory. These social forces might lead the theories to become more accurate, as theorists are forced to consider new evidence they had previously ignored. Or they might lead the theories to become less accurate, as theorists attempt to accommodate the objections and concerns of vocal agitators even at the cost of ignoring or discounting evidence. Either way, these changes are not the result of the social world being "on the move," but the result of the fact that what we think and communicate about the world is, in part, shaped by background theories and social forces that are, in the relevant sense, independent of what we are representing.

Putting all these elements together, then, we get that whether the social world is stable enough to sustain our successful knowledge of it in a particular case depends on a number of factors (see Figure 7.2).

(1) Success rate: the rate at which our epistemic activities allow us to converge on the truth. A higher rate means the world seems relatively stable.
(2) The degree of Taylor Instability: the rate at which the part of the social world we are describing changes for reasons that are independent of our epistemic projects (the right-side top-to-bottom arrow.) A higher rate of change means less stability.
(3) Labeling effects: the rate at which our epistemic activities cause the world to conform to our theories (the left-to-right arrow). A higher rate of conformity means more stability.
(4) The degree of Hacking Instability: the rate at which our epistemic activities cause the social world to change in ways that deviate from our theories (also the left-to-right arrow). A higher rate of deviation means less stability.

These factors dynamically interact (with others) to create stability or instability in each case. Hacking, like Taylor, suggests that instability is a cause for epistemic concern, but it's unclear that pessimism about knowledge of the social world is warranted. Why think that the rate of disconfirming change of the social world will outrun the mechanisms by which our theories couple to it? The considerations we have adduced thus far suggest that far from pessimism about the stability of the social world, we ought instead to consider the matter an open question, to be decided by the particular mixture of constructive forces that exist in each case.

3 Coupling Theory and World

In this section and the next, I offer a range of very general considerations that tend to support the stability of social categories. My aim in this is not to insist that culturally constructed categories must be stable, but to make clear that there are forces that support stability over time, and thus, it is not surprising that we can have knowledge of the social world that is real and useful.

3.1 Stabilizing the social world with cognitive constraints

On the picture of social reality that both Taylor and Hacking have, social reality is partially constituted by the intentional states of members of the social group, and these intentional states are themselves constituted by the concepts and conceptions of those members. Both worry especially that conceptual change and innovation will result in the instability of the social reality that our concepts structure, leading to failures of knowledge. However, this raises the question: how rapid and unbounded is conceptual change?

Dan Sperber (1996) has suggested that humans have innate predispositions to think about the world using some conceptions rather than others. Sperber offers an explicit and unapologetic evolutionary explanation for such predispositions: they are adaptations to thinking about the world in certain ways because those ways were helpful to us in adapting to our evolutionary environment. We can see this idea illustrated by returning to the example of essentialist thinking that I discussed in Chapter 1. There we considered evidence that conceptions of human racial groups were themselves partially caused by a human tendency towards construing some sorts of groups in an essentialist way. Cast in Sperberian terms, essentialist construals of biological or human groups amount to a cognitive predisposition to represent some groups as having essences and reasoning about them, as such. Crucially for our present argument is that if some version of the Sperberian hypothesis is right about the concepts that structure human social reality, then we ought to expect that these predispositions will buffer social reality from conceptual change that is at odds with those predispositions. Social institutions that seem more intuitive or "natural" will be more likely to be adopted and sustained. If this is true, then it suggests that certain aspects of certain intentional states may be "sticky," stabilizing the social reality they structure.[12]

None of this is to say that conceptual change or social transformation is not possible. (In Chapter 1 we considered several actual examples of deviation from essentialist thinking.) It is only to say that, given such predispositions, we might expect that human social institutions will be more likely to occupy these sticky sub-sets of the logically possible social arrangements. If so, then such cognitive dispositions ought to slow the rate at which social structures change, adding to the stability of the social reality that we attempt to track. Endorsement of such biases amounts to a deviation from pure social constructionist accounts of the representations that structure social roles, but it is perfectly consistent with a constructionist account of the categories such representations

---

[12] In a similar vein, a number of other evolutionary psychologists have argued that certain social institutions are structured by human cognitive tendencies toward certain sorts of social arrangements (Fiske 1991; Boyer and Petersen 2011), claims that, if they are true, would tend to stabilize aspects of social reality even as other forces transform it.

bring into being or causally structure. Such hybrid constructionism offers the category constructionist a mechanism with which to resist runaway instability of the social world.

3.2 Constructing categories and stabilizing categories

When we look, more specifically, at the mechanisms of category construction that we considered in Chapter 3, we find that many of them are such that they produce stability in the kinds in question.[13] Here, I briefly review telling examples.

3.2.1 INTENTIONAL ACTION

Consider, to begin with, the mechanisms of intentional action that are crucial to accounts of construction suggested by, for instance, Hacking, Appiah, and others. In these cases, category representations that are common knowledge in a population allow individuals to engage in predictable, coordinated behavior with others. Hacking's own account of multiple personality disorder has it that patients enact a widespread representation of multiple personality in response to features of their situation (distress or trauma) and in order to achieve certain social benefits (perhaps therapeutic care, recognition, exculpation for actions). In Chapter 5, I provided a model on which all this can happen while the patient continues to sincerely misexplain her own reasons for action. Close consideration of Hacking's own explanations of looping effects looks most readily to suggest that labeled persons may come to resemble their representation in a theory, leading the changes to the social world to result in confirming, rather than disconfirming, those representations. Work from social psychology like Carol Dweck's also suggests ways the content of putatively natural categories might produce self-fulfilling prophecies by presenting putative category members with information about their likely success or failure at certain sorts of endeavors, providing rational incentives that curtail their development of certain skills and traits.

That intentional mechanisms *can* give rise to stable regularities is well established by game theory. Such game theory models social facts as the outcome of the behavior of rational actors that seek to maximize their own utility, and it is an upshot of such models that rational actors' strategic choices can give rise to stable equilibria. It follows that if social roles coordinate the behavior of individuals in ways that produce such equilibria, then such roles can produce stable behavioral

---

[13] Murphy 2006, 267–9 argues for the possibility of stable socially constructed kinds in a similar vein. Cf. Mallon 2003.

regularities over time. For instance, if category representations produce Nash equilibria, situations in which no actor can unilaterally deviate from a choice without experiencing a worse outcome, then stable regularities could result from the rational, intentional choices of the persons involved (Mallon 2003; cf. Lewis 1969).

3.2.2 AUTOMATIC PROCESSES

Representations can systemically influence categorized persons via their influence on automatic psychological processes that mediate much human behavior. For instance, cases of stereotype threat suggest that representations of categories as being better or worse at certain activities may influence the behavior of category members in ways that seem to confirm the stereotypical content. While there is no single, widely accepted explanation for these effects, most models emphasize a substantial role for a range of automatically triggered processes (e.g. Schmader et al. 2008).

Similarly, representations may implicitly influence a community's behavior towards category members (as in cases of implicit bias) in ways that produce a regime of sometimes minute social rewards and punishments for stereotype deviant actions. And work on emotional "display rules" suggests that such rewards and punishments can themselves produce sophisticated automated, category-typical behavior (Ekman and Friesen 1972).

3.2.3 ENVIRONMENTAL CONSTRUCTION

Representations also guide changes to the environment that themselves, via feedback, can increasingly differentiate category members. Recall that part of the motivation to incorporate such environmental elements is to explain the obvious and lasting reality of some contemporary social categories. If we consider many contemporary United States cities, we find cities marked by sometimes very sharp residential segregation by race, segregation that also tracks other socio-economic variables like wealth, income, and education levels. While these differences have been produced in part by practices employing category representations, now that they exist, they need not disappear with any simple shift in concept use.

If this is right, then environmental construction not only produces category difference, it acts to stabilize and sustain conceptual and social practices that distinguish people. Not every constructionist category need be sustained by massive construction of the environment. But to the extent that a category is so sustained, we ought to expect it to be stabilized by its connections with the material environment over long periods of time.

3.3 Social construction and stability

The upshot of these considerations is that when we look at mechanisms by which representations might produce or partially produce human categories, they plausibly sometimes favor scenarios in which representations produce regularities in the world that, in turn, stabilize (by providing evidential support for) the representations. Such feedback loops are at the heart of the strongest case for believing in socially constructed categories. It follows that the social constructionist commitment to category construction need not be a commitment to categories that are fundamentally unstable, unable to support the acquisition of knowledge.

Indeed, to the extent constructed categories result from categorized persons coming to resemble the widely held theories put forward to describe them, we might expect constructed categories to be *more tightly coupled* to our representations of them than natural kinds are.

4 Stability and Human Natural History

While I have been arguing for the possibility of stable socially constructed categories, I have allowed that they might also be unstable for some of the reasons that Taylor and Hacking emphasize. And while I have presented evidence for mechanisms that would support such stability, this can seem so much cherry-picking of evidence, since we could as easily have sketched mechanisms or scenarios in which instability would have resulted. Thus, one might worry that, at best, what I have produced is an argument that stability could possibly obtain, sometimes. Given the manifest variability of culture—the obvious variation in culture and behavior that obtains both at the level of the individual and between groups—why should we take seriously the idea that culturally caused or constituted social roles might themselves sometimes be rather stable?

Further reason for believing in stability is provided by consideration of the role of social roles in human social life and human natural history. Call the entire system of social roles that characterizes a particular social milieu a *social ordering*. When we ask why we have this social ordering rather than that one it leads us to focus upon individual and group variation. Given this contrast, our social ordering seems arbitrary and contingent, and its changes over time make it seem unstable. But when we ask why a social ordering exists at all, the contrast is much different. Social orderings exist at a time because they represent the ongoing effort by groups of humans to adapt themselves to the natural and human environments around them. Given this functional role, the need for some social ordering with at least spatio-temporally local stability becomes compelling.

If broad, evolutionary considerations seem to favor the stability of social orderings, these considerations can serve as proposals of ultimate explanations for the proximal mechanisms that we have already sketched. They provide background considerations for why various psychological forces, some of them perhaps unconscious, might converge upon our behavior conforming to widely understood representations.

4.1 Human natural history and ultrasociality

Since splitting with chimpanzees (perhaps 5 to 7 million years ago), humans have undergone rapid evolution of body and brain, developing remarkable intelligence, range of linguistic expression, cultural capacities, and prosocial capabilities that have enabled them to live in norm-governed ways in larger and larger groups. At the same time, we have spread across the earth to establish ongoing communities in a vast range of environmental circumstances.

The standard story for what has made our rapid evolution possible is the coming into existence of certain selective feedback loops in which competition and cooperation *with other humans* has driven the accumulation of greater and greater capacities for language, for understanding other minds (Leslie 1994), for sentimental signaling (e.g. Kelly 2011), for culture (Richerson and Boyd 2005), and for regulation of oneself and others by social norms (Sripada and Stich 2006; Chudek and Henrich 2011; Boehm 2012; Kitcher 2011; Kurth 2015). Against this background of human natural history, we can say that for humans, evolutionary success has meant evolving capacities for efficiently coordinating and cooperating and competing with other humans (Henrich and McElreath 2003). Humans are profoundly prosocial, and concepts and norms that govern human categories figure as part of the broad set of cognitive mechanisms that we use to negotiate the social world—negotiate our relationships with other humans and groups of humans.

One general moral of this sort of reconstruction of human natural history is that it is not at all surprising to find humans to have an other-things-equal disposition to acquire and conform to local norms, perhaps enacting local, culturally given social roles. Because we cannot simply make our own society up on terms that favor our values or ourselves, and because the benefits of coordinating with others are so great, conformity to a role arrangement will in many cases be more locally worthwhile than opting out of a social arrangement all together. This has perhaps been *more* true throughout human evolutionary history than today; it is easy to imagine that ancestral human societies took a dim view of nonconformists. It follows from this line of thinking that those who were better able to reap the advantages of group life would have been favored by natural selection, and that these selection pressures may have shaped human psychology in important ways.

4.2 Tribal instincts for social roles?

If an evolutionary problem for humans is how to live successfully in norm-governed groups, how did we respond to it? One possibility is that we have become especially sensitive to membership in different social groups. We saw in Chapter 1 Francisco Gil-White's (2001a) suggestion that humans might have exapted biological cognition to pick out ethnic groups, understood as groups of humans who share culture, distinguish themselves symbolically, and follow distinct norms. Gil-White hypothesizes that essentialist thinking about human groups is adaptive because cooperating and coordinating with those who share the same norms as you do would be more efficient and safe than cooperating with out-group members (cf. Henrich and McElreath 2003).

Lots of the psychological evidence regarding group cognition fits with this general idea. It makes good sense, for example, of the psychological appreciation of mere distinction among "minimal groups" as these are plausibly an adaptation to the importance of social groups in human natural history (e.g. Chudek et al. 2013; Greene 2013; Richerson and Boyd 2005, chapter 6; Richerson and Henrich 2011; Kurzban et al. 2001; Pietraszewski et al. 2014; cf. Chapter 3, Section 2.1).

Similar considerations suggest the hypothesis that human cognition could be adapted to the existence of social roles within an extended community, where these take the form of physically or symbolically marked persons who have different properties and for whom different norms of interaction obtain.

Some have suggested as human groups became larger, a shift had to occur from cooperative and coordinative mechanisms that relied on altruistic moral sentiments (sympathy, compassion, a sense of fairness) to those that relied on a "norm psychology" that applied more abstract norms that did not depend upon the immediate occurrence of moral sentiments like empathy (e.g. Boehm 2012; Kitcher 2011). But larger groups bring other problems, including the problem of interacting with others with unshared background beliefs using only our folk "theory of mind." If in meeting each new person, we must infer the person's beliefs, motives, and possible actions, in order to engage in coordination and cooperation with them, it would presumably be at a substantial "computational cost."

A solution that seems to be employed in contemporary social life is the existence of norms and roles that limit the space of interactions with others, providing us with scripted forms of norm-governed interaction that obtain even with strangers. It is a feature of large contemporary communities to have norms and scripts that concern only some groups or roles. Such norms, scripts, and groups allow us to bracket the myriad particular mental states of particular people, treating them primarily as occupants of certain positions or roles. We can then interact with social role occupants in routinized ways—ways governed not by explicit agreement, but by common knowledge of roles. On this view, role-specific norms are a sort of culturally specified heuristic for negotiating life in a cultural milieu—they cut through decision space, limiting the consideration of features of persons for certain purposes (Bermúdez 2003; Maibom 2007).

Like differences among ethnic groups, differences among social roles are often marked by special, symbolic dress and other markers signaling membership in the social role, and we take these signals to indicate the normatively acceptable schemes of interaction. So a different speculation than Gil-White's—one that requires at least as much cognitively but suggests richer cultural consequences—is to suggest not only adaptations to track ethnic groups but adaptations that accommodate us to within-group social roles.[14] Like between-group membership, within-group role differences are often marked symbolically, often with the apparent purpose of "enhancing the signal," making role membership more conspicuous. For instance, men and women plausibly typically dress differently in American culture in part because dress signals social role. It is a potential cue for the sorts of interactions that are and are not desired or reasonable or normatively appropriate. Similarly, contemporary occupational roles often offer clear categories and scripts for interaction, smoothing social exchange, cooperation, and coordination, marked in many cases with indicators of dress that are conspicuous signals of social role membership.

4.3 The imperative of social life and social role stability

Evolutionary considerations suggest a plausible ultimate explanation for social stability rooted in the imperatives of social life: we need to coordinate and cooperate with others, and we typically have little meaningful option to "opt out" of the cultural milieu in which we find ourselves. It makes sense that we would have a range of capacities that would enable us to find a role within the social ordering in which we find ourselves, and these capacities would lend

---

[14] Within-group social roles might have their origin in the merger of smaller ethnic groups into larger human groups in which sub-group-specific norms continued to obtain (by occupation, tribe, caste, or so on). In such a circumstance, we might expect as a by-product of Gil-White's essentialism about ethnic groups to be essentialism about sub-groups within a culture.

themselves towards *signaling that we belong to a role* and *understanding the roles signaled by others* instead of toward undermining and destabilizing those roles. The social ordering fulfills an essential social function of organizing the coordination and cooperation of large groups of relatively unrelated and unfamiliar people. It is thus plausible that, in addition to the myriad factors that lead to its transformation over time, there are other factors that tend to hold it in place.

5 Stable Enough Is Stable Enough

We now have reasons to believe in the possibility of stable-enough socially constructed categories. Critics of such stability often emphasize the instability of social kinds because such kinds are not rooted in biological facts. Such categories have, in the words of the critical sociologist Stuart Hall, "no guarantees in Nature" (1996, 166). But having no guarantees in nature is not the same as having no guarantees at all. A particular social setting, structured by a conception of a human kind, may be guarantee enough for a particular social role to create stable explanatorily relevant property-cluster kinds that support our epistemic projects over periods of time that are interesting to us.

But stable-enough categories are only one requirement for the successful acquisition of knowledge of a category. We also require the capacity to co-refer to the category over time, and constructionists have special challenges in accommodating co-reference. It is to these I now turn.

8

Achieving Reference

Consider these two different questions about a human category:

What is widely *believed* about the explanatory mechanism or kind that underlies the category by the population (some of) whose members putatively instantiate the category?

What is the *real* character of the explanatory mechanism or kind underlying the category?

Suppose (too simply) that there are only two sorts of answer to the first question, two sorts of beliefs: *beliefs in natural kinds* and *beliefs in social kinds*. And let us say also that there are only two possible types of explanatory mechanism underlying the phenomena associated with a human category term: natural kinds and social kinds. Then we have four possibilities:

	Community believes it's a	Really is a	Possible examples
1. Overt	social kind	social kind	Mayor of Indianapolis, Member of Parliament
2. Covert	natural kind	social kind	Race, gender, multiple personality disorder
3. Transparent	natural kind	natural kind	pregnant, tuberculosis
4. Reduction	social kind	natural kind	schizophrenic, alcoholic, depressed, epileptic[1]

To this, we might add the possibility of failed reference. This generates two additional, skeptical possibilities:

---

[1] These categories might once have been considered disorders of character, perhaps brought on by social arrangements and choices, but now are more often considered disorders of brain or biology.

	Community believes it's a	Really is a	Possible examples
5. Skepticism	social kind	nothing	Yuppie, hipster
6. Skepticism	natural kind	nothing	fairy, witch

We have been concerned especially with line 2, covert constructions. Covert category constructionists share two commitments that are important here. First, they are focused upon the possibility of cases where there is a widespread, but mistaken, explanation of features of an apparent kind in terms of an underlying, putatively natural kind. Second, they hold that human category terms like "race" (and racial terms) and "sex" (and sexual terms) and "homosexuality" and "perversion" successfully refer to real categories that are caused or constituted by our social and linguistic practices: they are "discursive formations" or "social constructions" or, as I call them here, *social role kinds*.

It is far from clear that these two commitments are consistent. The sort of mismatch between widespread belief and underlying nature that constructionists take to hold between what is widely believed about a referent and what is really the case is often taken to indicate a failure of reference, a situation in which our term does not refer at all. If we think of terms like "fairies," or "witches," or "unicorns," for example, it is tempting to think that these terms don't refer to anything in the actual world since there are no things in the actual world of which the beliefs people associate with these terms are true. These considerations give rise to the question: how can we be sure that, even where constructionists are right about the sources of category difference, human kind terms pick out social constructions (line 2) rather than nothing at all (line 6)? Call this *the mismatch problem*.

Mismatch arguments drive some theorists to deny the existence of race. As we noted in Chapter 1 (Section 1.4), many racial theorists have suggested that racial essentialism is criterial for the race concept. On these views, "race" just means a sort of human group whose members have a distinguishing essence. However, the existence of racial essences (as of biological essences more generally) is now widely rejected by contemporary biological theorists. The pairing of these two theses—the conceptual claim about race and the biological claim about racial essences—generates the mismatch between theory and world that is at the heart of much skepticism about the existence of race. Indeed, *the* core argument for racial skepticism in the recent philosophical literature is:

Semantic Premise: It is part of the definition of "race" that it picks out a sort of human group that is characterized by an essence.

Metaphysical Premise: There are no essences.
Semantic Conclusion: "Race" does not refer.
Metaphysical Conclusion: Race does not exist.

And most every philosophical skeptic of the existence of race does so, at least in part, on the basis of such a mismatch argument. They conclude that our folk racial discourse is simply mistaken, and that there are no races.

But a mismatch produces trouble not only for biological realists about race, but also for any social constructionist about human kinds that shares the two commitments. By the skeptic's lights, it seems that the constructionist is simply making a mistake. Whatever constructed kinds could be, they are *very different* from the ordinary (perhaps biological or essentialist) understandings of the kind terms that are associated with them. It follows that if satisfaction of these ordinary understandings are what fix the reference of human kind terms, then the kind terms do not refer—not to social constructions, not to anything.

The mismatch problem is closely related to another, classic question in the philosophy of science: how could successive scientific theories co-refer? Realists typically have it that theories of a domain successfully approximate the truth, and scientific realists hold that more recent scientific theories approximate the truth more successfully than previous ones, converging on the truth over time. But because new theories may make dramatically different claims than old theories, they can seem to be talking about different things. Kuhn famously defended such semantic incommensurability, and, as we noted in Chapter 6 (Section 2.2.1), this has often been taken to imply a failure of co-reference. On such an anti-realist reading of Kuhn successive theories do not converge on truth about a subject over time. They instead change the subject. Avoiding this result requires an account of reference that supports co-reference. Call this *the problem of co-reference*.

The problem of co-reference, like the mismatch problem, afflicts constructionists as well. Continuing with the example of race, recall Fordham and Ogbu's controversial though influential claim that American blacks' lower academic performance compared with white peers is the result of a black culture that devalues excelling at school as "acting white" (Chapter 3, Section 1.3). This claim has resulted in a great deal of debate. Tyson et al. (2005) directed a North Carolina study that found no difference in school attitudes between white and black students in most schools examined. Fryer and Torelli's (2010) analysis of data from high schools across North America suggest there may still be something to the idea, but, as Harris (2011) argues, this explanation has limited explanatory scope.

Importantly, if racial conceptions are rapidly evolving, for example in response to the first black U.S. President, or increasing awareness of unfair treatment of different races by law enforcement, or increasing scientific knowledge of human natural history, or findings about human genetic diversity, or awareness of implicit racial bias, then it becomes unclear whether "black" or "white" used about students in a Washington, D.C. high school in the mid-1980s *could* mean the same thing as "black" or "white" in North Carolina, or high schools across North America in the 2000s. Unless we take it to be *possible* that the "black" and "white" of various studies pick out a common subject matter in somewhat different cultural contexts, contexts where the descriptions associated with the terms are altered, sometimes substantially, it is unclear how they can play the role they do in social scientific discourse.

Mismatch and co-reference are not the only problems that face social constructionists in connection with reference. Consider, for instance, worries about *trivial reference*. In ensuring that (some) human kind terms refer to social constructions rather than nothing, constructionists have to avoid guaranteeing reference in every case, so that every term (even "cyclops" and "phlogiston") refers to something (making lines 5 and 6 empty). Another problem concerns *collateral reference*. Social roles of some sort surround not only covert constructions, but also real natural kinds. If terms for putative natural kinds can refer to social roles in the covert case (line 2), why don't natural kind terms actually refer to social roles all the time, making the Transparent or Reduction cases empty (lines 3 and 4)? As we have emphasized, there could also be cases of merely *partial construction* where social role terms pick out an entity that is in part cultural construction and in part natural kind. Why don't they always occur (again making lines 3 and 4 empty)? Addressing this range of problems requires getting a better grip on how and when terms of massively mistaken theories might refer.

Fortunately, the mismatch problem and the problem of co-reference have a widely recognized solution that is now part of the basic understanding of realism in many areas of philosophy of science: reject the claim that the beliefs or understandings or descriptions associated with terms for human kinds are what fix the referents of those terms, and rely instead on some *external* account of the reference relation, for example a *causal-historical* account of reference of the sort that Saul Kripke and Hilary Putnam famously advocated. Crucially, on such a causal-historical theory, a term may refer to some kind or stuff present at the introduction of the term that entirely fails to satisfy the beliefs or descriptions that contemporary people associate with the term (Kripke 1972/1980; Putnam 1975).

Because of its capacity to avoid skepticism from mismatches and ensure co-reference, causal historical theories of reference have now been invoked in a range of domains, and a number of philosophers have suggested social constructionists might also avail themselves of this approach to reference, mimicking the traditional realist response (e.g. Boyd 1992; Haslanger 2003, 2005, 2012; Mallon 2003). Here, the social constructionist suggests that while a term like "race" or "sex" or "homosexual" is widely believed to pick out a natural kind, it really (according to the external facts that determine reference) picks out a social construction. As long as we hold that such reference is achieved by some external relation, one that does not require the truth of beliefs associated with a description, it seems that reference to a kind can be permissive of a mismatch with widely held theories of a kind, and co-reference can be permissive of widely varying conceptions of the kind. In what follows, I develop this idea, focusing on answering the mismatch problem (though I return at the end to consider other problems of reference).

I have two aims here. The first is to make clear that the constructionist appropriation of causal historical accounts of reference is beset by difficulties that do not attend the natural kind theorists' use of such accounts. In particular, since constructionists apparently hold that our linguistic, conceptual, or social practices constitute a human category that comes to be instantiated, then it seems to follow that at the moment that a term or social practice is introduced, *there was no referent there to refer to*, or at least *there is no referent of the right sort*, and therefore *there is no referent to causally historically ground our own use of human category terms*. It is thus unclear how human category terms could come to refer to social constructions. Without such an understanding, the coherence of a broad and influential range of constructionist claims is in doubt. My second aim is to argue that these difficulties can be answered, and that the social constructionist view—that terms for apparently natural human kinds refer to some sort of social construction about which there is massive error—could be correct.

I begin, in Section 1, with some background assumptions about the nature of the search for a correct reference relation. In Section 2, I consider the difficulties confronting the constructionist attempt to appropriate the causal-historical account of reference, and I suggest that these difficulties can be answered by appeal to a successful account of reference *switching*, which in turn requires an account of *grounding*. "Grounding" in this chapter picks out the relationship that a term acquires to its referent during a naming event or "baptism" that allows later uses of the term to refer to the referent. (It thus does not pick out the metaphysical grounding relation that has been the subject of so much recent philosophical work; e.g. Schaffer 2009.) In Section 3, I argue that the category constructionist again faces special difficulties offering an account of grounding. In Section 4, I turn to begin to develop an alternative account that allows for grounding in cases of massive error. In Section 5, I close by briefly sketching a constructionist solution to achieving reference, and I reconsider other problems of reference that beset the constructionist.

1 Which Theory of Reference Is Right?

Skeptics about the existence of kinds typically get support from massively false theories associated with the terms for those kinds by assuming some sort of "descriptivism" about reference (Lycan 1988; Stich 1996). Traditional descriptivist accounts of reference hold that:

> D1. Competent speakers associate a description with a term t. This description specifies a set of properties.
>
> D2. An object is the referent of t if and only if it uniquely or best satisfies the description associated with it.

Descriptivist accounts of reference have proved crucial in skeptical arguments across a range of philosophical domains, including eliminativism in the philosophy of mind (Churchland 1981; Stich 1983, cf. 1996), and as we just noted, debates over the reality of race. When reference-fixing descriptions are not satisfied, the descriptivist skeptic concludes the terms do not refer, and that the relevant entities—e.g. beliefs or races—do not exist.

In contrast, on a causal-historical approach to reference, terms refer via some sort of causal-historical connection (mediated, perhaps, by one's community or by a community of experts) with a referent. Causal historical accounts hold that:

> C1. A term t is introduced into a linguistic community for the purpose of referring to a particular thing (e.g. a person or a property). The term continues to refer to that thing as long as its uses are linked to the thing via an appropriate causal chain of successive users: every user of the term acquired it from another user, who acquired it in turn from someone else, and so on, back to the first user who introduced the term.
>
> C2. Speakers may associate descriptions with terms. But after the term is introduced, the associated description does not play any role in the fixation of the referent. On such an externalist approach, our terms refer via some extramental connection to the world rather than by means of our beliefs about the referent.

The success of the constructionist appropriation of externalism about reference looks to hinge upon the particular details of such an account for the case of concern. But a prior question is whether causal-historical accounts of reference are, in fact, true of human kind terms. While this view commands a wide following, others employ descriptivist approaches (e.g. Lewis 1970, 1972; Jackson 1998), or hybrid approaches (e.g. Evans 1973). Who is right? How could we know?

I have considered these questions at length elsewhere (Machery et al. 2004; Mallon 2007b; Mallon et al. 2009; Nichols et al. forthcoming), and my own view is that folk semantics supports more than one way of construing the reference of kind terms. On this view, token uses of kind terms can be ambiguous between different semantic construals—for example, ambiguous between descriptivist and causal historical construals of the referent of a term.[2]

But in what follows I put aside questions about the correct reference relation, and I simply assume that a causal-historical theory will be correct for connecting human kind terms to social constructions if we can specify the details of this connection in a plausible way. This assumption is appropriate since my argument here concerns precisely whether this condition can be discharged. If some other account of reference is correct, then category constructionists will not be able to appeal to causal-historical accounts of reference to respond to the mismatch problem and other problems of reference.

2 The Missing Referent and the Wrong Kind of Referent

Causal-historical theorists typically insist that a term attaches to a kind via some sort of external relation that is established via some initial "dubbing" or "baptism" or grounding event. Michael Devitt and Kim Sterelny explain that,

As a first approximation, the grounding of a natural kind term includes both an ostensive component and a "nature" component. In a paradigm case, such a term is introduced into the language by ostensive contact with samples of the kind. Thus, "tiger" is introduced by causal contact with sample tigers and "gold" by causal contact with samples of gold. The extension of the term is then all those objects, or all those examples of stuff, that are of the same kind as the ostensively given samples, that share the underlying essential nature of the samples.   (1999, 88–9)

The historical sequence on this canonical account goes something like this:

time	word–world relation
t_0:	Before grounding. "T" does not refer.
t_1:	Grounding event: "I/We hereby name *that kind* a 'T'!" "T" refers to a kind, K.
t_{now}:	Now, via a chain of reference borrowing: "T" still refers to K.

---

[2] This suggestion has been defended by Philip Kitcher in the philosophy of science (Kitcher 1978, 1993; Stanford and Kitcher 2000), and it has also received recent support from Shaun Nichols, Ángel Pinillos, and myself (forthcoming). In a series of experiments, we shift the construal of kind terms between descriptivist and externalist construals. On this picture there may be more than one semantic reference relation that may be correctly applied to a particular utterance.

But, as I noted, this canonical account looks to fail for constructed kinds. To see this, suppose that we could travel back in time to t_0, before the initial practice of using terms to distinguish a covertly constructed kind. It seems as though the constructionist should insist that, at t_0, *no kind existed*. The "nature component" is missing. For example, thoroughgoing constructionists about race or homosexuality believe that *before* the first person to use racial terms or the term "homosexual," there was no natural, biobehavioral kind to be ostended, and by hypothesis, no socially constructed kind as well. But if no kind existed at t_0, then it seems that no kind existed to be the target of ostensive or indexical naming at t_1 either. The constructionist account of categories and the causal historical account of reference seem to entail that human category terms like "race" or "sex" would not refer at t_1. (In contrast, the natural kind theorist does not have this problem for the natural kind theorist can posit a natural kind, however poorly understood, as the referent at t_1.) Call this *the problem of the missing referent*.

In response to the problem of the missing referent, it seems that the constructionist has two options. One is to insist that, against my (perhaps oversimple) reconstruction of the social constructionist position, there really is a referent to be the target of grounding—viz. there is a referent at t_0 or at t_1. The second is to suggest that although human kind terms failed to refer to social constructions at t_1, they have since *switched* referents, and so, whatever was initially the case, they now refer to social constructions. Let us consider each option in turn.

2.1 There is a referent

What kind of referent might, on the constructionists' account, ground the initial use of human category terms? Here I consider two options. First, I consider the possibility that there is a *thin* natural kind at t_0 that can ground reference at t_1. Then, I consider the alternate possibility that an *institutional kind* referent is present at t_1—that, in effect, the same activities that introduce the use of a kind term simultaneously constitute its referent.

2.1.1 AT TIME T_0: THIN NATURAL KINDS

Have I been too quick to insist that social constructionists need to insist that there is no natural kind present to be the referent at the initial introduction of a kind term? The social constructionist can surely allow that construction begins by attaching significance to a thin natural kind, where "thin" simply means a kind that may be objective and natural but is nonetheless explanatorily weak. It plays little or no role in the explanation of other properties. Constructionists can say that race is objective but only "skin deep," for instance, or that sexual categories, which can be of reproductive significance, have no systematic psychological significance. Once we allow that such thin natural kinds exist, it seems plausible to imagine a primordial observation of some difference or differences among people or peoples. For instance, it might begin with observations that:

(1) Some encountered group of humans shares different behavioral dispositions and physical features than one's own community.
(Consider races or ethnicities.)
(2) Some group of humans within one's community has systematically different behavioral dispositions or capacities or physical features.
(Consider sex categories, or categories of sexual orientation, medical categories, or occupations.)
(3) Members of one's community, or some members, occasionally have systematically different behavioral dispositions or coordinated suites of behaviors.
(Consider emotional categories, or personality traits, or transient mental illness categories.)

Suppose that these observations lead people to the essentializing thought:

People with properties $p_1, p_2 \ldots p_n$ share membership in some natural category NK and we hereby dub this kind "T."

Because the thin natural kind theorist posits the presence of a thin natural kind, such an act of categorization might, in fact, be true. That is, if we consider, say, the readily observable differences that are associated with races or sexes, or the differences of sexual dispositions that are associated with sexual orientations, then we might—even on a constructionist account—find a systematic though not inductively rich natural explanatory mechanism that might underwrite an initial grounding.

The thin natural kind constructionist can attempt to answer the problem of the missing referent by asserting that both before and at the time a term was introduced (at t_0 and at t_1), there existed a thin natural kind. The sequence is thus:

time	word–world relation
t_0:	Before the grounding. There is a thin natural kind, TNK.
t_1:	Grounding event: "I/We hereby name *that kind* [ostending an instance of TNK] a 'T'!" "T" refers to TNK.
t_{now}:	Now, via a chain of reference borrowing: "T" still refers to TNK.

Prima facie, the claim that there are such thin natural kinds is a very plausible thing to say about many human kinds, and it looks to be already part of some constructionist positions. Considering again the case of race, here the dominant constructionist position is that racial groups are not distinguished by membership in any inductively rich natural kinds, but even here most everyone allows that *some* biological features cluster together and correlate with some racial classifications.

Supposing, then, we allow that natural kind terms initially pick out thin natural kinds. While that would solve the problem of the missing referent, the category constructionist will not be happy to say that such thin natural kinds are "all there is to" or "give the essence of" contemporary categories like race, gender, or sexual dispositions. On such a view these categories would not be social constructs; they would be thin natural kinds. A thin natural kind is the wrong kind of referent for the constructionist to appeal to underwrite a constructionist account of some contemporary category.

It follows that even if thin natural kinds served as the referent at the introduction of a human kind term, we need some account of "reference switching" by which a term changes from one referent (a thin natural kind) to another (a socially constructed category). I return to this suggestion in Section 2.2.

2.1.2 AT T_1: INSTITUTIONAL KINDS

When one reads constructionist claims that social kinds like race or homosexuality are "invented" or "made up," it is tempting to interpret the view as claiming that human kind terms refer to something they bring into existence at the very time that they are posited. (Recall Foucault's claim that "the psychological, psychiatric, medical category of homosexuality was constituted from the moment it was characterized" (1978, 43).) This suggests the possibility that homosexuality or other human kind terms refer to something that is created at t_1, just as the term itself comes into use.

John Searle (1995, 2010) has offered an account of institutional kinds that understands such social institutions as collective "impositions of function," as the collective acceptance that: "X counts as Y in context C." Searle argues that this very basic collective acceptance can be used to give an analysis of a wide range of institutional aspects of social reality. And in fact, it seems that we do sometimes allow that social reality can come into being instantaneously in something very like this way in the case of social institutions.

Suppose a few of us get together to talk about Northern Cardinals. We have so much fun, we decide to form a club centered on our interest in bird watching. We agree to call it "The Midwest Birdwatchers Club" or "The MBC," and we elect a president to lead us. The good times roll. It seems plausible to think that the club comes into existence at about the same time that we name it. And similarly for the office of president of the MBC. And it also seems that people might refer to the club, over time, via some causal-historical connection to the entity created at this initial dubbing.

Institutional kind terms like "Member of Parliament," "U.S. President," or "Member of the Midwest Birdwatchers Club" seem to label kinds that come into existence at about the same time as the terms themselves, solely in virtue of the shared intention of the group to act as if the kind exists. And in these cases, terms for kinds seem to refer to the overt, institutional kinds that they are said to.[3] In such cases, the naming sequence might go like this:

time	word–world relation
t_0:	Before the grounding.
t_1:	Grounding event: A community collectively accepts that: "X counts as T in conditions C." "T" refers to an institutional kind, IK.[4]
t_{now}:	"T" continues to refer to IK.

If this is right for overt institutional social roles, could it be the case for covert category constructions as well?

No. As I suggested in Chapter 2 (Section 3.3), covert constructed kinds are not institutional kinds in this sense. Consider that, for covert kinds, the terms for them are widely believed to pick out natural kinds that are independent of the act of introducing the term. That is what it is for the construction to be covert. That is what gives rise to the mismatch problem. But it also gives rise to a number of important disanalogies with overt, institutional kinds.

We can note first, the epistemic disanalogies. As Edouard Machery (2014) has recently noted, on Searle's account of institutions, institutions emerge from collective acceptance of status functions. But on the constructionist account, differential treatment emerges from a population of individuals accepting that a category of human is natural. To see how different this is, we might note Amie Thomasson's (2003) argument that because Searle's functions are imposed by "we intentions" or collective acceptances by a population, members of that population will have a kind of a priori knowledge regarding the imposed properties:

---

[3] While I take it that such synchronic creation *can* be the case, I do not take it that all institutional kinds or social roles are produced in such an act of creation. There could be a gap between the time in which a term begins being used, and the time in which a genuine institution comes to exist.

[4] Things are more complicated than this for Searle now allows that there can be "freestanding Y terms" in which there is an institution, but no X to which this status is assigned (see Searle 2010, 20f.). I ignore these complications here.

we cannot conceive of investigations into the nature of our own institutional kinds as completely a matter of substantive and fallible discovery. Whereas natural kinds (on a realist view) can exist even if no one knows of their existence or any facts about their nature, institutional kinds do not exist independently of our knowing something about them. Similarly, whereas, in the case of natural kinds, any substantive principles any individual or group accepts regarding the nature of the kind can turn out to be wrong, in the case of institutional kinds those principles we accept regarding sufficient conditions for the existence of these entities must be true. We are guaranteed freedom from complete ignorance and are preserved from error in many of our beliefs regarding the nature of institutional entities precisely because the principles accepted play a stipulative role in constituting the nature of the kind. (589–90)

This kind of a priori knowledge is at odds with the a posteriori knowledge we take ourselves to have of the natural world, and it is at odds with the hypothesized mismatch that obtains with constructed kinds.

Perhaps the clearest way to show that these are different kinds of kinds is to emphasize their ontic differences. Most importantly, the stability of institutional kinds and the stability of covert kinds are sub-served by different sorts of mechanism. Acceptance of status functions grows out of human cooperative tendencies and is conditional upon the behavior of others. By way of example, consider a regularity that prefers women for certain sorts of employment, say nursing or teaching. A person might participate in such a regularity because there is an institution around here such that women count as better nurses or teachers than men. But if the collective acceptance of such a status function ends, one's own reasons for participating in it would as well.

In contrast, suppose that one really believes that women are better at certain tasks—believes that such superiority is among the typical effects of natural sex difference. In that case—the sort of case covert constructionists are concerned with—one's behavior is sustained by beliefs about the (putatively) natural features of the category itself, or about beliefs as to what others will do, given that they believe in the (putatively) natural features of the category itself. Thus, for covert social roles, but not institutional kinds, belief that the category is natural stabilizes the behavior surrounding the category, and the behavior that is conditioned on that behavior, and so on. This stabilizing mechanism distinguishes constructed categories from conventional ones.[5] Once a category is widely believed to be natural, the explanation for the continued, stable existence of the category is no longer rooted in our collective intentions, but in our beliefs about

---

[5] Searle (1995, 2010) suggests that institutional facts of the sort he is interested in can be mistaken as natural. The considerations reviewed here suggest that he is wrong.

the natural world. The mechanism by which the category continues as a powerful organizing factor in social life changes.[6]

If this is right, then even if we allow that institutional kinds are present (because created with) the introduction of a human kind term, such an institutional kind fails to be the sort of kind that constructionists insist underlies the contemporary use of covert kind terms. As with the proposal of thin natural kinds, we can still imagine that reference was initially grounded in an institutional kind but has switched to some covertly constructed category. Thus, here again, we find ourselves in need of an account of how the reference of kind terms could have switched.[7]

2.2 There is a referent after t_1: grounding and regrounding

It remains that the constructionist may allow that while a term for a constructed category initially referred to nothing, or to a thin natural kind or an institutional kind, it has since switched or become *regrounded* to refer to some sort of social construct.[8]

Gareth Evans's (1973) illustrates the possibility of switching by considering the proper name "Madagascar." "Madagascar," he notes, originally referred to a region of the African continent itself (apparently to what is now called "Mogadishu"), but was mistakenly applied by Marco Polo to the large island off of the southwest coast of Africa that it now names. Since "Madagascar" now names the island Madagascar, the referent must have switched at some point from part of the mainland to the island itself. Applying these thoughts to constructed kinds, we get a temporal sequence that looks like this:

time	word–world relation
t_0:	Before grounding. There may be thin natural kinds or no kinds.
t_1:	First grounding event: "I/we hereby name *that kind* a 'T'!"
But:	"T" refers to nothing.
	Or "T" may refer to a thin natural kind, TNK.
	Or "T" may refer to an institutional kind, IK.

---

[6] Recall Plato's use of the "myth of the metals" to stabilize the social order of *The Republic*. Cf. Mallon forthcoming.

[7] It follows from this argument that if an entire community followed the constructionist (as, say, Glasgow (2009) urges for conceptions of race) in abandoning false natural kind explanations of category-typical properties that the constructed kind would cease to exist (though, of course, it might be replaced by another, institutional kind).

[8] Many theorists would want to add, additionally, that where the process of switching is gradual, situations of *partial reference* have occurred (Field 1973). I ignore these complications here for simplicity.

t_2:   A covert social role kind SRK comes to exist.
t_3:   Reference switches. "T" is regrounded. "T" refers to SRK.
t_{now}:   Now (via a chain of reference borrowing from those at t_3) "T" refers to SRK.

Since Evans's groundbreaking discussion, switching has become part of the standard quiver of possibilities to handle problems with externalist reference, and it figures prominently in externalist debates in the philosophy of mind (e.g. Burge 1988; Tye 1998). Social constructionists who want to interpret human kind terms as referring to a social role kind like SRK can pursue a similar tack, allowing that at their introduction, such terms may have failed to refer to a construction, but insisting at some point after time t_2 the referent *switched* to refer to the social construction.

But there are difficulties with a straightforward application of Evans's idea here. "Madagascar" came to name Madagascar because it was used by Polo to name the island, and subsequent mapmakers picked up this new use. There was no mismatch between Polo's intended use and the current referent. However, in the case of terms for social kinds that are the targets of social constructionist claims, terms like "race" (or "black," "white," "Asian," "Latino," etc.) or "sex" (or "male" or "female"), or "homosexual" (or "gay," "lesbian," "queer," etc.), there is (by constructionist hypothesis) a considerable mismatch between what is believed about the kinds and the underlying reality of the kinds. Indeed, these terms have not until recently, and then primarily by social theorists, been used with the explicit intention to refer to sorts of social roles.

It follows that constructionists need an account of (re)grounding, on which a term can come to refer to a social role kind despite massive error about the character of the kind. Unfortunately, the most prominent account of the grounding relation is itself descriptivist, and so the social constructionist again runs into something like the mismatch problem, posing special problems endemic to the constructionist position.

3 The Qua Problem

Suppose Adam encounters a thing that he wishes to give a name. He points to it and says "I dub that an *elephant*!" Suppose everyone else learns the kind name "elephant" from Adam, and that even people who cannot tell an elephant from a hippopotamus successfully refer to the kind via connection with Adam's initial grounding. The causal historical theory treats Adam as having grounded the term in the kind elephant, and others as having borrowed reference from him.

But how, exactly, did Adam ground the name "elephant" in elephant kind? Adam ostended something, but the animal Adam saw belongs to many natural kinds (animal, mammal, proboscidean, warm-blooded, herbivore, elephant, mother). And it also falls into many other categories as well (big thing, cute thing, hairy thing, gray thing, loud thing, scary thing) (Devitt 1981, 60ff.; Devitt and Sterelny 1999, 90ff; Sterelny 1983, 116). Exactly which of these potential referents did Adam name "elephant"? And in virtue of what? This is the *qua* problem.

By way of answering the qua problem, Devitt and Sterelny, suggest that "something about the mental state of the grounder must determine which putative nature of the sample is the one relevant to grounding" (1999, 91).[9] They go on to hypothesize that the grounder of a term associates,

consciously or unconsciously, with that term first some description that in effect classifies the term as a natural kind term; second, some descriptions that determine which nature of the sample is relevant to the reference of the term. (92)

Thus, on their view, Adam names elephant kind (rather than something else) with "elephant" in part in virtue of his beliefs in using the name, and it is in virtue of the object that Adam ostends satisfying some description that Adam associates with it that Adam is able to ground the term "elephant" in elephant kind rather than not something else.

The qua problem has thus pushed many who, like Devitt and Sterelny, seem convinced of the basic soundness of Kripke's and Putnam's arguments in favor of a causal historical account of reference toward some hybrid account of the reference of kind terms, one that uses a description to achieve an initial grounding and causal-historical elements to account for reference borrowing (see, e.g., Devitt 1981; Devitt and Sterelny 1999; Stanford and Kitcher 2000).

Crucially, the descriptions used in grounding cannot be very general specifications, like "the natural kind" or "the thing that is ostended," for these specifications will not disambiguate the relevant kind. For example, they will not distinguish elephant kind from the many other kinds to which the elephant belongs. Following Devitt and Sterelny, Thomasson (2007) similarly argues that solving the qua problem requires that terms be associated with certain very general "frame-level" application and co-application conditions; "'frame-level application conditions'," she writes, "since they involve conditions that are *conceptually* relevant to whether or not reference is established" (39). And (though they ultimately qualify this view), P. Kyle Stanford and Philip Kitcher

---

[9] Devitt (1981) develops such a view for proper names.

initially imagine the grounding involves an appeal to "minute parts" of the referent (2000, 113). Call this use of a rich description to solving the qua problem by appeal to a robust description an *opulent baptism*.

It follows from this account that cases of radical error, cases like Hilary Putnam's famous case in which we discover the things we call "cats" are actually Martian robots, are cases where our frame-level application conditions (in which we understood our referent as, say, a natural kind of animal) are violated (Putnam 1975), and thus Putnam is describing a case where our term "cat" was *never successfully grounded* in robot-cat-kind. It would be a case of *failed* reference.

For similar reasons, opulent frame-level application conditions look to exclude constructed social kinds. Since frame-level conditions are hypothesized to involve categorical concepts (like "natural kind" or "biological kind" or ones involving "minute parts") that are not true of constructed kinds (by constructionist hypothesis), it apparently follows that these too will be cases of failed groundings and therefore failed reference. The mismatch problem thus returns to sabotage constructionist efforts to use description satisfaction in grounding, just as mismatches sabotaged successful reference to covert social constructions on traditional descriptivist accounts.

Because solving the qua problem with an opulent baptism places limits on the sort of errors that are compatible with successfully referring in the grounding situation, it deviates somewhat from the spirit of a pure causal-historical approach (though insisting on true frame-level conditions for grounders is consistent with the thought that some or all *later* speakers—speakers that refer via reference borrowing—might have a completely false description). This has led some critics to insist that this approach must be mistaken. Laura Schroeter (2004), for instance, writes:

Aristotle and his contemporaries apparently thought that fire was a substance; we think it's a process. We disagree about the basic metaphysical nature of fire, but according to common sense interpretive intuitions we're still thinking about the same phenomenon. Once again, it's not just our semantic intuitions which are at stake, but common sense epistemology. Empirical inquiry, we want to say, revealed the truth about fire—the very phenomenon we were thinking about all along. (2004, 439)

Schroeter's criticisms of theories of reference that employ a descriptivist solution to the *qua* problem will resonate with those who insist on the compatibility of very substantial error with successful reference. But, how, then, are such terms grounded? How can the covert constructionist and other defenders of reference-cum-radical-error solve the qua problem?

4 Solving the Qua Problem with the Explicans

We can retain the dominant descriptive approach to grounding if we can find a solution to the qua problem that does not employ the sort of opulent baptism that gives rise to mismatches. But is it possible to specify "frame-level application conditions" *generally enough* that a natural kind term could pick out a socially constructed mechanism, but still *specifically enough* to enable the description to determine the right kind?

The obvious place to look for such criteria is to refocus upon the *cause* or *the explanation* of the phenomena that lead us to posit an underlying kind—ideas that are already implicit in existing accounts of grounding. Devitt and Sterelny, for example, suggest the following procedure:

> People group samples together into natural kinds on the basis of the samples' observed characteristics. They observe what the samples look like, feel like, and so on. They observe how they behave and infer that they have certain causal powers. At some level, then, people "think of" the samples under certain descriptions—perhaps, "cause of O" where O are the observed characteristics and powers—and as a result apply the natural kind term to them. It is this mental activity that determines which underlying nature of the samples is the relevant one to a grounding. (1999, 92)

If we can follow Devitt and Sterelny and conceive of the grounding as specifying the "kind that is the cause of certain features of the sample" then we can perhaps show the constructionist is no worse off than her biobehavioralist opponent: both appeal to surprising causes of kind-typical properties.

In order to do this, we first need an understanding of an underlying kind that allows us to understand cultural constructions as possible candidates. One way to make this idea explicit is to exploit Richard Boyd's influential causally homeostatic property-cluster kind view, as I did in Chapter 3. On Boyd's account, the kinds that support induction in many sciences can be characterized as a cluster of more or less coinstantiated properties together with some mechanism that explains their co-occurrence. Crucially, Boyd's account is neutral about the nature of the properties that co-occur and of the mechanisms that are responsible, paving the way for constructionist accounts of such properties and mechanisms (Boyd 1988, 1999b; Mallon 2003).

But there remains the worry that demand for the explicans of, say, "observed characteristics and powers," is not by itself sufficient to distinguish a socially constructed kind from other kinds, that we have not yet solved the qua problem. Stanford and Kitcher, for example, emphasize the thought that attention to the common constituent of a set of samples (and excluding a set of foils) still does not uniquely pick out a single natural kind (2000, 110). Their own solution is to offer a further refinement in two parts. First, they allow that a grounder of a term has "a very rough description of the inner constitution she wants to pick out." And second, they imagine her as fixing upon "that feature of the minute parts (whatever it is and whatever they are) that is a common part of the total causes of each of the distinct ascertainable properties" (113). In effect, they attribute to her a sophisticated description under which she can "partition the total cause" into the relevant underlying mechanism (114). However, neither of Stanford and Kitcher's refinements can be straightforwardly applied to the constructionists' problem.

First, for the reasons just considered, the social constructionist is going to be unhappy with reference to the requirements of "inner constitutions," "inner structure," and "minute parts," for, according to the constructionist, the kind that explains category difference is not produced by distinctive "inner structure" or "minute parts," but, rather, at least in substantial part, by distinctive social relations. In effect, even these very general specifications of the grounding description run afoul of the constructionists' need to achieve reference while allowing for error.

Second, at least some of the core cases we have been imagining do not fit perfectly well into the "causal partitioning" strategy they evoke. As we just conceded in our discussion of the possibility of thin natural kinds, race and sex, for instance, both plausibly involve a folk theory that connects some innate, biological features of the body (e.g. skin color) with various other traits that might include behavioral capacities, dispositions, and relational facts. In the case of race, while the folk and historical scientific theorists may have suspected that different typical traits have a common cause (perhaps an underlying, intrinsic essence), the contemporary social constructionist imagines a social practice that explains many differential features of race while allowing that other bodily traits (for instance, those that are used as indicators in ascription) might have a separate, biological explanation. It follows that the social role will not be a "common cause" of all the characteristic features of paradigmatic instances.

Consider these problems in turn, beginning with the description sketch for the underlying kind. I have little doubt that grounders of human kind terms often conceive of the kind they are attempting to pick out as an element of "inner constitution" or of "minute parts." As I argued in Chapter 1, research into folk psychological essentialism suggests that the folk conceive of many kinds, including biological kinds and racial kinds, as having underlying essences that both explain category boundaries and category-typical properties. Thus, I see no reason to doubt that the folk think of their groundings of natural kind terms in just the way Stanford and Kitcher's discussion suggests, as involving mechanisms that are, for instance, "minute" and "inside." But while this may be part of the way would-be grounders conceive of the kinds they label, the qua problem needs to be solved without appeal to *where* the explanatory mechanism is (inside or outside), appealing instead *only to* causal partitioning to find the relevant explanatory kind.

To say this might seem to leave the constructionist worse off than her biobehavioral-kind opponents, for opponents can more tightly constrain grounding descriptions than constructionists, offering a better solution to the qua problem. However, as Schroeter emphasized above, constructionists are not the only ones who have reason to appeal to a more liberal account of grounding descriptions to answer the qua problem.

Consider biological kinds. Folk conceptions of biological kinds plausibly involve, initially at least, essentialist assumptions about those kinds (Keil 1989; Gelman 2003). It is plausible to think that these assumptions are psychologically present at grounding. However, contemporary theoretical biology widely rejects the existence of such essences. As Sterelny and Griffiths write: "In reality...there is no such thing as the 'genetic essence' of a species...Diversity is normal... uniform populations in the natural world are unusual" (1999, 7). Instead, many contemporary biologists view species members as sharing distinctive, kind-typical properties in part in virtue of population-level barriers to genetic flow, in effect, replacing an *intrinsic* construal of kindhood with a *relational* view (though see Devitt 2008, 2010). Ernst Mayr, in defending his classic biological species concept, wrote: "It is...irrelevant and misleading to define species in an essentialistic way because the species is not defined by intrinsic, but by relational properties" (1984, 535). If this relational construal is correct, then many *biological kind* terms are characterized by mismatches between contemporary beliefs about their referents and their real natures, and also that this was so at the time the terms were first grounded. If we wish (in Schroeter-esque fashion) to think that contemporary utterances of species-kind terms co-refer with past utterances that were grounded prior to the contemporary antiessentialist revolution in theoretical biology, we must allow that grounding even a natural kind term need not necessarily appeal to internal constitutions or "minute parts." And the point is not limited to just biological kinds. As Muhammad Ali Khalidi remarks, "it is not clear that there are any macroscopic kinds that can be characterized solely in terms of microstructural properties" (2013, 37).

Such examples suggest that it is not only the constructionist that must seek a non-opulent baptism for grounding, but many others as well. (Stanford and Kitcher themselves, in further elucidating their proposal, come to emphasize the need for an account of some natural kinds that need not involve reference to intrinsic structures (120ff).) The constructionist is thus in the good company of at least many contemporary theoretical biologists. While people widely believe (and scientists may once have widely believed) that kind terms picked out instances characterized by internal properties that disposed kind members to certain traits, including certain sorts of behaviors, many now believe that there may be no such internal properties. Constructionists can join theoretical biologists in holding that our terms continue to refer despite the failure of traditional understandings of what they refer to.

Return to the second problem, the problem of the absence of a common causal component in the explanation of the core features of members of our sample. Again, the comparison with species seems relevant. Consider a set of species-typical traits for, say, the striped skunk. Consider both the characteristic fur patterns and its signature defensive perfume. While folk biological essentialism might have it that the explanation of both traits resides in a common cause, an essence, contemporary biologists would presumably explain the development of these traits by appeal to distinct though overlapping environmental and genetic resources.

Suppose that the DNA sequence that (partially) controls the characteristic hair pattern is distinct from the DNA sequence that (partially) controls the stink. Then the total cause of *stripedness* will overlap with the total cause of *smelliness*, but the intersection of the two total causes will not include the controlling DNA sequences (since each sequence, by hypothesis, plays no part in the "total cause" of the other trait). The worry, generalized, is that if we take the intersection of mechanisms that make up these overlapping total causes of the characteristic traits of our samples, we are likely to *miss out* on the particular mechanisms that control the distinctive expression of the characteristic traits.

This shows what is already implied in Boyd's account of property clusters: we do not want merely a common explanatory component of the individual properties, but of their reliable co-occurrence and co-instantiation—of why the skunk-typical properties co-occur in skunks. Once that is part of what is to be explained, we will have to appeal to the various mechanisms by which those different DNA sequences come to co-occur and co-express in *Mephitis mephitis*. If this is right, then the category constructionist can avail herself of the same appeal: we neither want an explanation of the biological traits nor of the (ex hypothesei) constructed traits, but of their co-occurrence in particular places and time. According to the social constructionist, the social role explains this co-occurrence.

No doubt this account can be further refined. But I have tried to show that, with regard to the *qua* problem, the social constructionist is no worse off than the natural kind theorist. Both should, at least sometimes, want to accommodate Schroeter's demand for the possibility of co-reference among wildly different theories. And a plausible strategy for doing so is by appeal to the cause or explanation of some ostended samples. The example of biological kinds suggests that we should allow that external relations, and not just internal "minute parts," might constitute a natural kind. And this example also illustrates that it is not merely a set of characteristic features, but their co-occurrence that needs explanation.

5 Achieving Reference: Taking Stock

It is time to take stock of where we are. I began the chapter by noting that the mistakes that are endemic to folk understandings of human kinds, on the constructionist picture, threaten the ability for terms of those kinds to refer to anything at all, and we then noted the familiar idea, that could be availed of by constructionists, that reference might obtain independently of the content of such folk understandings, via some external, causal historical relation.

However, we also noted that externalist accounts of the reference relation still have to be grounded in kinds, and that the social constructionist metaphysical picture implies the absence of any kinds, or any kinds of the right sort, at the times that terms for those kinds first came into use. In none of these cases, we suggested, was the original use of the term co-referential with the constructionists' account of the contemporary use of human kind terms to pick out some sort of social role kind.

A natural thought was to invoke the familiar idea of reference switching, claiming, in effect, that human category terms have switched from nothing, or thin natural kinds, or institutions, to refer to social constructs. On the account offered, category constructionists can hold that this switch occurred at the point at which the important properties of the kind—the properties that we hope are explained by appeal to a kind—came to be those that were caused by the existence of the social role in the community. The model we have of achieving reference through construction is thus one that involves causal looping that produces changes in both the extension and intension of the concept. Because this process occurs over time, the exact location of the point at which the term clearly refers to an extension that was partially caused or constituted by earlier uses of the term will be somewhat vague, but that needn't keep us from recognizing points that are clearly before and after this transition.

The picture of externalist reference also suggests resolutions to the other concerns about reference that we raised at the beginning of this chapter: the problems of co-reference, trivial reference, collateral reference, and partial construction:

Co-reference: Common reference among two terms figuring in distinct theories of human categories is achieved whenever the two terms are grounded in the same kind. This can happen when: two different theories reference borrow from the same grounding event, or when each borrows from a grounding event that picks out the same kind.[10]

Trivial Reference: Terms for nonreferring kinds like "fairies" and "cyclops" do not refer because there is not, and never was, an entity (and certainly not a social role) that grounded the use of these terms by, e.g., explaining the core features associated with these kinds. "Witches" is harder. Are contemporary, self-identified "witches" actually witches? Here the answer is unclear. Exactly what phenomena need to be explained by appeal to a kind for it to count as witchhood? Insofar as we come to use "witches" to pick out a contemporary, self-identified group of people who share certain cultural practices, key texts, social networks, and so forth, then it might become grounded in those groups, and thereafter refer to them. But achieving this reference switch, from nothing to something, is not trivial.

Collateral Reference: Terms for natural kind terms should refer to those kinds and not to the social roles that come along with common knowledge representations of a kind. On the theory here, a term is grounded to the kind that, inter alia, explains a set of features in the samples of the kind that ground it. So long as these features are not explained by the accompanying social role, the kind term won't (in whole or in part) pick out the social role. To the extent that they are, it may.

Partial Construction and Reference: What about cases of partial construction? In Chapter 3 I approvingly quoted Richard Boyd, who wrote: "there should be kinds and categories whose definitions combine naturalistic and conventional features in quite complex ways" (1992, 140). We have now seen a range of sorts of kinds that may be in play: what I called covertly constructed kinds (sketched in Chapters 2 and 3), institutional kinds (discussed briefly in Chapter 3 and in this chapter), and natural kinds. Now we are in a position to suggest: reference to entities that combine different sorts of mechanisms can be appropriate to the extent that these mechanisms figure in the common causal explanation of the features whose description grounds the term.

---

[10] This is a sufficient, but not a necessary condition for co-reference.

Using a causal-historical approach, the constructionist can accommodate reference to kinds that are pure covert or institutional constructions, those that combine covert and institutional elements, and those that incorporate natural kinds of increasing causal relevance.

With this picture of reference in place, we now return to where we began. We can imagine that as ideas about race, or sex, or homosexuality, or multiple personality disorder came to have causal power, they differentiated putative category members from others in more and more robust ways. These aspects then became noticed, theorized about, and ultimately became a crucial part of the phenomena to be explained by our use of the respective human kind terms. The reference of such terms ultimately shifted from whatever it was before to the social role kinds (or combined kinds) that use of the terms brought into being. This story is plausible, and it shows how some provocative constructionist claims of category invention could be true.

It vindicates something like Hacking's (1986) "dynamic nominalism" on which our practices of thinking about, describing, and differentially treating persons can give rise to the very kinds that those practices describe and refer to. And it also suggests the possibility of a different sort of conceptual break (promised at the end of Chapter 1): If our concepts and beliefs are individuated partly by their referents, and if the introduction of new terminology and social practices can constitute a new referent *for that very terminology*, then once the introduced terms shift to refer to these novel referents, people can think thoughts, utter words, and perform actions with new, never before available meanings.

PART III

Conclusion

9
Alternatives and Implications

Entrenched social roles offer a "how possibly" model for the covert construction of real human kinds. At the core of the account are social roles that come to produce systematic causal effects on persons and the environment in a particular social milieu. The homeostatic property cluster that results can become the referent of the category term, sustaining various epistemic projects regarding the category.

I have emphasized that interest in the character of human categories emerges from a range of sources, and that among them are explanation and a concern with social regulation, often out of concern for social justice. From such diverse sources, a wide range of work on human categories has developed in recent decades. I conclude here by positioning my own entrenched social roles account of covert construction among some alternatives for understanding human categories. This is not an exhaustive survey of the possible positions, but something more like a look around the theoretical neighborhood. I close by considering the role of constructionist revelation in social change.

1 Nonconstructionist Alternatives: Natural Kinds and Skepticism

Covert social constructionism about human categories most obviously contrasts with views on which human categories are caused by or constituted by or metaphysically grounded in natural kinds—especially biological or medical kinds—that do not necessarily involve human mental states, decisions, culture, or social practices.

If covert social constructionists are right, we are sometimes mistaken about the character of the mechanisms that cause or constitute important human categories, and this ignorance threatens to undermine our social-regulative projects and our self-understandings. I have suggested that the constructionists' conception could be realized by common knowledge representations of a category becoming entrenched, producing causally differentiated human categories (Chapters 2 and 3). The error regarding the character of such categories is compounded wherever they become falsely essentialized (Chapter 1) and wherever they serve as a basis for mistaken explanations of oneself and others (Chapters 4 and 5). All this amounts to an interpretation of the social construction of human categories that places explanatory power at the center of our interest in the categories. In contrast with much constructionist scholarship, I have engaged this discussion at a rather high degree of generality, but also with attention to a range of experimental evidence. Whether or to what extent the ideas and models I have sketched are actually realized for real human categories is something that is still to be shown.

On my account, the covert constructionist also contrasts with skepticism—the view that the relevant category does not really exist. Skepticism is tempting in nearly every case of covert construction because, by hypothesis, covert construction involves serious error about the properties that underlie a category—error that is easy to interpret as evidence that our attempts to pick out a category with a human category term have failed. I have parried skepticism by appealing to the causal importance of the human categories that constructionists are interested in (Chapters 2-3), and by providing a model for how our terms for such categories might refer to them in cases of serious error (Chapter 8).

The account I offer, if applied to race, thus differs from Joshua Glasgow's (2009) *substitutionism* in part because it holds the constructionist can sustain coreference with an ordinary nonconstructionist account of human categories it hopes to displace. Glasgow thinks we ought to use something much like racial terms to pick out the social constructions that our racial thought and practices produce, but he holds that ordinary racial terms do not do so because they entail commitment to an "adequate biological basis" that does not, in fact, exist (2009, 114). In contrast, I have argued that if human categories like race or gender are covert social constructions with significant causal power, then it could be reasonable to consider them as the referents of our ordinary racial and gender category terms.

I say "reasonable" rather than "required" because I think that at best the metaphysics and semantics can array the ontological options on the table, but that choosing among them—choosing exactly what we should say about the existence of and how we should talk about the character of human categories—is an all-things-considered judgment that is underdetermined by the metaphysical and semantic considerations, and partially determined by political and social context (Mallon 2006, 2009; Nichols et al. forthcoming; cf. Gannett 2010).

2 Other Paths to Construction: Justice-Driven Metaphysics

I have framed the question for social constructionists as a search for an account of social construction that could plausibly support the causal power such categories apparently have in social life and social science. Thus, entrenched social roles are a possible answer to questions posed by what we could call *explanation-driven* metaphysics. However, many theorists approach socially important human categories animated primarily by concerns with social justice, and in particular concern with the role that social categories can play in systematic stratification of persons based on membership. For instance, in an influential discussion Sally Haslanger (2000/2012) has proposed accounts of gender and race that are not accounts of the meanings or referents of ordinary gender and race terms, but proposals for how we ought to understand gender and racial categories in pursuit of certain critical feminist and antiracist projects. For example, Haslanger suggests that:

S *functions as a woman* in context C iff$_{df}$

(i) S is observed or imagined in C to have certain bodily features presumed to be evidence of a female's biological role in reproduction;
(ii) that S has these features marks S within the background ideology of C as someone who ought to occupy certain kinds of social position that are in fact subordinate (and so motivates and justifies S's occupying such a position); and
(iii) the fact that S satisfies (i) and (ii) plays a role in S's systematic subordination in C, that is, *along some dimension*, S's social position in C is oppressive, and S's satisfying (i) and (ii) plays a role in that dimension of subordination. (2000/2012, 235)

Haslanger intends this definition (as well as her related accounts of "functioning as a man" and of racialization) as theoretical tools that can guide thought and action in social progress. In some ways, Haslanger's account resembles a social role account of the sort I have defended. However, our accounts differ fundamentally in the essential reference Haslanger's account makes to the normative concepts of "oppression" and "subordination" (and elsewhere, of "privilege"). This is no mistake, for, by Haslanger's lights, it is these normative concepts that mark the domain of relevance for progressive social theory. In contrast, they have played little role in my explanation-driven approach.

I agree that explicitly normative, politically driven accounts of human categories have an important role to play in shaping our social, ethical, and political discourse. We can acknowledge this—acknowledge the importance of what Haslanger is trying to do—while allowing that there are a range of legitimate and valuable interests and methods with which one might approach discussion of human categories, a point that Haslanger herself makes persuasively (2012, chapter 6).

Once we allow that multiple projects drive concern with human categories, we can ask how the results of these projects—explanation-driven and *justice-driven* metaphysics—are connected. In principle, they need not be in conflict. Different approaches to carving up the world could arrive at the same account regarding the metaphysics of a human category: what we should represent them as being for the sake of social progress could turn out to be exactly how we represent them in our best explanations. (Something like this idea seems to motivate some constructionists' desire for revelation.) But even where explanation- and justice-driven motivations lead to different accounts of human categories, they may not conflict, for they are answers to different questions that involve satisfaction of different constraints. Alternative approaches to human categories may simply latch onto, or illuminate varying aspects of the world for their varying purposes.

Along the way, I have alluded to a number of ways in which the entrenched social roles account of human categories that I have developed offers vindication for some normative concerns surrounding the representation of human categories (e.g. by illuminating the production of moral hazard (Chapter 4), by figuring in mistaken self-explanations, and by producing failures of agency (Chapter 5)). As this suggests, I hold that my entrenched social roles account illuminates certain normative issues, and it is driven in part by the view that our normative ends are not served by leaving our explanatory ends behind. We have to understand how things might work to understand how they do work. And we should understand how they do work if we hope to move them towards how they should work.

3 Varieties of Constructionist Metaphysics

The social role account of covert construction that I have provided is based on the idea of social roles that are structured by the representations of human categories and, over time, by the causal effects of such representations. As such, it is a particular interpretation of how to be a social constructionist about human categories, one that perhaps most readily echoes the Foucaultian idea of a "discursive formation" or Hacking's "dynamic nominalist" approach to human categories. The account I have offered is, however, only one way to develop social constructionist themes, and it may be worth contrasting with a few others.

Perhaps the simplest constructionist idea is just the *nominalist* idea that the boundaries of our representations are entirely determined by convention, and pick out no real borders or boundaries "out there" in the world. On the view I have defended, even properties whose boundaries are determined entirely by human convention are objective enough features of the world, because humans and our minds and concepts and social practices are natural, real things (Chapter 6). But I also developed an account of human categories in which social roles go beyond simple nominal distinctions to result in socially and causally significant kinds. The causal importance of the resulting categories also plays a role in explaining how human category terms for covert constructions could come to refer to the constructions themselves (an account unavailable to the simple nominalist account of those categories). The resulting entrenched social role view thus does more than a nominalist account because it provides a metaphysical explanation of the causal power of constructed human categories and of the partial success of our (everyday, social scientific, and theoretic) discourse about social reality. But it is also committed to more as well: it is committed to the world being arranged in particular ways.

An alternative constructionist approach understands social construction in accord with recent work on overt social or institutional reality. A fruitful body of philosophical work in recent decades has provided illuminating accounts of conventions, group cooperation, and institutions (e.g. Lewis 1969; Gilbert 1989; Searle 1995, 2010), and it is tempting to understand covertly constructed social reality by appeal to these detailed models. This temptation is heightened by the frequent use by covert constructionists of overt institutions as illustrations of how human practices can produce social reality (e.g. Griffiths 1997; Zack 1999, 2002; Mallon 2004; Haslanger 2012).

Work on the metaphysics of conventions and institutions is important, and my own account incorporates overt institutions as a mechanism that enhances the causal significance and stability of covert constructions. However, on the dominant account of overt social reality, conventions and institutions are sustained by dispositions that are conditional upon the preferences or cooperation of others, and I have argued that covert constructions are instead conditional upon a mistaken view of the nature of the categories in question (Chapter 3 and Chapter 8). Because the mechanisms that sustain the categories are different, the entrenched social role account offered here is a different kind of kind from the institutional account (cf. Mallon forthcoming).

In other respects, my account remains neutral on central questions regarding the metaphysics of social categories. We can see this by considering Ásta Sveinsdóttir's (2013) recent defense of what she calls a "conferralist" account of social construction, one that understands a category C as socially constructed in virtue of being conferred on a person *p* by a social practice. Ásta imagines our category representations to assign persons to social roles that bring with them certain "constraints and enablements" (724)—a view that fits easily alongside the social role account I offer in this book. But what distinguishes Ásta's account is her insistence on the centrality of actually conferring a property in a context.

Ásta holds that what makes a token *t* a member of a category C is that *t* is actually judged to be a C and treated as a C. For instance, on a conferralist construal, a pitch in a baseball game is not a strike in virtue of some physical properties that it has (for example, having a certain trajectory), but in virtue of having been judged to be a strike by the umpire. For Ásta, the umpire's call confers *strikehood* upon the pitch. Similarly, a person is a *woman* in virtue of being judged to be a woman in a context.

Both Ásta and I hold that there are social practices ($k_1, k_2, \ldots k_n$) that ascribe sex to persons on the basis of their apparent possession of putatively natural properties ($a_1, a_2, \ldots a_n$). These ascriptive practices ($k_1, k_2, \ldots k_n$) also have systematic consequences ($r_1, r_2, \ldots r_n$). We can then distinguish the folk or ordinary view, Ásta's view, and my own view by the properties that constitute or ground the kind:

> The ordinary or "folk" view holds that membership in a category C consists in possession of putatively natural properties ($a_1, a_2, \ldots a_n$).
>
> Ásta's view holds that membership in a category C consists in being actually classified as a "C" by the manifestation of ascriptive practices ($k_1, k_2, \ldots k_n$).
>
> My view holds that membership in a category C consists in possession of some sufficient cluster of properties that may include putatively natural properties ($a_1, a_2, \ldots a_n$), being classified by practices ($k_1, k_2, \ldots k_n$), or the consequences of such ascriptive practices ($r_1, r_2, \ldots r_n$).

Thus understood, Ásta's conferralist view might be seen as a more specific development of, or a particular sort of, social role view of the sort I have offered. (However, I have been especially focused upon covert constructions and Ásta's account cross-cuts both covert and overt or institutional accounts of categories.) In contrast with Ásta, my account is neutral with regard to the constituting or metaphysical grounding properties for a category in part because I have been developing a toolset rather than a specific account of specific categories.

In fact, I hold that the relevant homeostatic property-cluster possession of which constitutes membership in a particular human category is to be determined a posteriori by investigation into the features that explain whatever phenomena ground use of the term. In this vein, as I consider the construction of social categories like race and gender, it seems important to me not to exclude in advance the possibility of partial construction, that is, the possibility that some actually natural properties figure alongside social roles and their consequences in grounding some categories (cf. Gannett 2010).

In addition, on the basis of what we already know, I am also skeptical that grounding categories in ascriptive practices alone (and *a fortiori* in actually exercised ascriptive practices) is sufficient to explain causally powerful social categories like race and gender, for the explanatory power of these categories may necessarily involve features that are the consequence of material transformation of our worlds. (We can see this difference by asking, as I did Chapter 3: if the practices of ascription disappeared, would the categories also disappear simultaneously?)

But these are partially empirical questions, and our evidence about what does and doesn't constitute a human category may change. The entrenched social role account here is neutral among a wide range of accounts of the exact properties that constitute or ground the kind.

4 Nonessentialism, Generalization, and the Boundaries of Kinds

The emphasis on explanation and realism in the present work may seem to run afoul of a different sort of tendency in social theory, a ubiquitous tendency toward nonessentialism about social categories. By "nonessentialism," here, I mean the view that categories or kinds do not share a common essence (rather than the view that an individual's being a member of a particular kind is essential to that individual's identity—cf. Witt 2011). I have emphasized (in Chapter 1 and again in Chapter 8) my understanding of constructionist categories and biological categories alike is nonessentialist.

Nonetheless, in contrast to much work on social categories that seeks to qualify or undermine the possibility of generalizations about category members, the present focus on the causal power of categories may instead be seen as trying to make space precisely for such generalizations. I have denied nominalism and skepticism about human categories, and understanding social categories as causally powerful underwrites generalizations. Generalizing, in turn, brings in its wake the risk of pernicious generalizing that can harm individuals and groups in a range of ways. Negotiating a path between the assertion of causally significant categories on the one hand, and the amelioration of the risks and harms of pernicious generalizing on the other, is a central problem of contemporary work on the metaphysics of socially important categories like race and gender (e.g. Alcoff 2006).

I think the first step in addressing this problem is to appreciate that there is no immaculate solution to it. We can make some progress by insisting that our category representations be true—by subjecting them to empirical scrutiny. However, even after we do so, there can be a tradeoff between the benefits and costs of various ways of representing human categories. We saw one set of reasons for this in Chapter 4: even true representations of human categories can modify our inferences and attitudes towards category members in ways that may have harmful consequences. It is also the case that even explanatorily strong generalizations may limn some people as exceptions or "outliers" in ways that could be harmful to them. For instance, representing heterosexuality as statistically the most common sexual orientation may result in reflexive "heteronormativity"—the adoption of norms and standards that assume heterosexuality—and thereby fail to acknowledge and accommodate the needs and interests of lesbian, gay, bisexual, and asexual persons. But rather than thinking that there is some careful formula by which we can both have our representations of human categories pick out real differences, but also never have these representations give rise to harmful consequences, we ought to appreciate that risks come hand in hand with representation, and that the costs and benefits of employing particular representations must be considered and evaluated in context and in dialogue.

Still, the causally homeostatic property-cluster view that I defend does offer amelioration of some harms of generalization. Because the account I have offered is a property-cluster kinds view, it is not essentialist in the traditional sense: it does not entail that there is some simple set of properties possession of which is necessary and sufficient for membership in a category.[1] Rather, the account is compatible with rich variations among category members even with respect to possession of category-typical properties.

In addition, once we allow that the proper boundaries of social categories are revealed a posteriori in the course of an investigation, we can articulate the underlying mechanisms and causal power of a category as an ongoing response

---

[1] In this way, it is a metaphysical counterpart to "family resemblance" understandings of concepts for human categories (Nicholson 1994, 1999; Stoljar 1995; Davidson and Smith 1999; Wittgenstein 1958; Cf. Mallon 2007a).

to our explanatory needs and the growth of our empirical and moral understanding. It follows that our representations of human categories are capable of progressively sophisticated articulation in light of counterevidence and counterexample. In particular, overgeneralized and too-weak characterizations of categories can be modified as we seek levels of generality appropriate to our explanatory aims. For similar reasons, single-dimensional analyses of human categories like race or sex can, where appropriate, give way to intersectional analyses of categories where it is membership in an intersecting category that drives explanatory power.[2]

5 Covert Construction, Revelation, and Collective Action

I close by considering, more generally, the role of covert constructionist work in bringing about social progress. I begin with the observation that even when we accept a covert category constructionist account of a category, we can continue to feel powerless to change the category itself. Rather than revelation leading to revolution, we can instead feel rather like a Martian anthropologist standing outside the social order that we think that we have come to understand. Why should this be?

One problem is that the correct beliefs of an individual or small group are not themselves sufficient to transform a social role produced by more widespread, common knowledge representations of a category. But changing widespread social practices and material arrangements that construct categories often requires collective action, and famously, collective action problems are hard to solve. Participation in collective action can be individually costly and may not succeed. If it does succeed, the benefits of collective action may be distributed more broadly than to just the participants. Suppose overturning gender social norms requires collective action by women and men to act in countertypical ways. In this case, each individual considering action faces this dilemma of cooperation:

> This cooperative social movement may or may not succeed. If it does not succeed, I will have paid the cost of action without receiving the benefits. If it does succeed, it will likely have done so whether or not I participate. In that case, I can gain the benefits without paying the cost.

---

[2] Where this occurs, the co-reference of correct social theories and folk terms for human categories may break down, perhaps replaced by situations of partial reference (Field 1973).

The uncertainty of success and the possibility of freeriding in such situations make successful collective action very difficult to sustain. Call this the *collective action* problem.

In cases of covert social construction, cooperation is even more difficult to achieve because there is an *additional* barrier to achieving collective action: the widespread belief that the state of affairs is the product of natural (perhaps biological) mechanisms. For covert category constructionists, mistaken belief in the naturalness of a category can be one of the mechanisms by which the category is sustained. Call this *the reification problem*. It follows that where such covert construction exists, widespread recognition of its existence undermines a central mechanism that sustains construction of a category, shifting our perception of our own roles from observers to possible actors in the creation or sustenance of these categories (cf. Griffiths 1997, chapter 6).

The existence of the collective action problem already shows why revelation is not sufficient for social change, for social transformation requires more than undermining reification. But the collective action problem and reification problem are actually mutually exacerbating. The reification problem makes the collective action problem worse because representations of category-typical behaviors as natural can lead subjects to reduce their belief in the possibility of successful action. The perception (whether or not true) that natural regularities will be robust in the face of collective behavior to alter them can lend credence to doubts that collective action can succeed. But it's also true that the collective action problem can exacerbate reification since failures to achieve collective action for whatever reason may be interpreted as further evidence for the naturalness of role-sustained regularities and for the impossibility of successful change.

Individual recognition of the possibility, or even of the actuality, of covert construction is only a start in bringing about social change. Social transformation requires ensuring cooperation in the face of the possibility of failure and of freeriding, but it also requires changing widespread attitudes that category regularities are an inevitable consequence of membership in a natural kind. A better understanding of potential mechanisms of covert social construction can be of help in this latter task, and the sort of work I have undertaken in this book is a modest contribution to this end.

Acknowledgments

My thinking about the many of the topics in this book germinated in conversations with, and then a dissertation under, Stephen Stich at Rutgers University. It was with Steve that I first began to try to articulate the philosophical and empirical import of social constructionist claims, and his own philosophical work and philosophical adventurousness has remained an inspiration in everything I have done.

I was also grateful during those years for the respite from teaching provided by a Charlotte Newcombe Dissertation Fellowship from the Woodrow Wilson Foundation. My development of some of the ideas in this book continued at the University of Utah and during a Research Assistant Professorship at Hong Kong University. The preparation of the present book manuscript began during a Laurence S. Rockefeller Visiting Professorship at Princeton University in 2005-6, and it was subsequently aided by a Sexuality and Gender Studies Research Award for Faculty at the University of Utah as well as sabbatical support from the University of Utah and later from Washington University in St. Louis. I was also helped along the way by a Faculty Fellowship from the University of Utah and an American Council of Learned Societies Fellowship in 2009-10.

Then and in the intervening years, I also learned a great deal from a community of scholars whose influence in person or through their work has informed my thinking and this book in pervasive ways. Especially important early on were Kwame Anthony Appiah, Richard Boyd, Paul Griffiths, Ian Hacking, Sally Haslanger, Charles Mills, and Naomi Zack; each of them showed me that a certain kind of philosophical exploration of constructionist ideas was possible and interesting. A partial list of others must include Linda Alcoff, Robin Andreasen, John Doris, Stephen Downes, Luc Faucher, Jorge Garcia, Tamar Gendler, Joshua Glasgow, David Theo Goldberg, Dan Kelly, Edouard Machery, Howard McGary, Dominic Murphy, Shaun Nichols, Lucius Outlaw, Michael Root, Tommie Shelby, Edward Stein, Ronald Sundstrom, and Paul Taylor. Members of the Moral Psychology Research Group have also been a neverending source of challenge, information, and inspiration in this as with all my work.

A large number of others—too many to list here, and certainly too many to remember!—have earned my deep appreciation for their specific comments and suggestions regarding the ideas and in some cases portions of the text that appears here. Some of these include: Chrisoula Andreou, Eric Brown, Carl Craver, Andreas De Block, Anne Eaton, Frankie Egan, Doug Jones, Muhammad Ali Khalidi, Stephen Laurence, Brian Locke, Tori McGeer, Aaron Meskin, Elijah Millgram, Elizabeth O'Neill, Alan Patten, Philip Pettit, Dan Robins, Jenny Saul, Frederick Schmitt, Dave Shoemaker, Walter Sinnott-Armstrong, Tamler Sommers, Cindy Stark, Mariam Thalos, Manuel Vargas, Chris Weigel, Jonathan Weinberg, Alison Wylie, and audiences at Arizona, Arizona State, Duke, KU Leuven, Princeton, Rutgers, St. Louis University, Texas Tech, Tulane, University of Houston, University of Missouri-St. Louis, University of San Francisco, University of Sheffield, Utah Valley University, and Western Michigan. I add to these thanks to the students in Dan Kelly and Daniel Smith's Fall 2014 Seminar at Purdue University and also to the students in my own Fall 2014 graduate seminar at Washington University in St. Louis: Michael Carver, Dylan Doherty, Maria Doulatova, Cameron Evans, Mark Povich, Rick Shang, A. J. Van Westen, Rachel Williams, and Tom Wysocki for much helpful discussion and criticism. I also benefited from helpful discussions about feminist metaphysics with Lisa Cagle, Abigail Klaasen, and Christiane Merritt.

My thanks also to Peter Momtchiloff at OUP for his persistence, patience, and optimism regarding my book project. Thanks also to the anonymous readers for OUP and OUP's terrific production team who have all helped make this a better book.

John Doris, Dan Kelly, Charlie Kurth, Shaun Nichols, and Anya Plutynski all deserve special acknowledgment for generously reading and providing commentary and criticism on an earlier draft of the entire manuscript.

All this excellent discussion, advice, and feedback has made this a much better book. It leaves me with the abiding wish that it was better still. Of course, the flaws that remain are my own.

This project also would have died long ago if not for the support of friends and family. Steve Stich has been not only a philosophical inspiration, but a personal one too. On several occasions, his encouragement was crucial in moving this project towards completion, and I am grateful for his confidence and his wise advice. John Doris has long been a source of friendship and steadying counsel, never more so than since I joined him at Washington University in 2011. I am also thankful to Shaun Nichols for his insistent encouragement over the years and for his apparently sincere claim that he wanted to read this book. Above all, I am indebted to my family: my mother and father, my children, and especially to my wife to whom this book is dedicated.

In writing this book, I drew on ideas, and in many cases text, from previously published work, and I gratefully acknowledge permission from their publishers to reuse this material here.

Portions of the introduction draw upon:

Mallon, R. (2007). "A field guide to social construction." *Philosophy Compass* 2(1): 93–108 published by John Wiley and Sons. © Blackwell Publishing 2006.

Chapter 1 reuses and revises material from:

Mallon, R. (2013a). "Was *Race* thinking invented in the modern West?" *Studies in History and Philosophy of Science* 44: 77–88 © 2012 Elsevier Ltd., with permission from Elsevier.

The text of Chapter 1 also draws upon:

Mallon, R. (2010). "Sources of racialism." *Journal of Social Philosophy* 41(3): 272–92, published by John Wiley and Sons. © 2010 Wiley Periodicals, Inc.

Portions of Chapters 2, 3, and 7 draw upon:

Mallon, R. (2003). Social construction, social roles and stability. In *Socializing Metaphysics*. F. Schmitt. Lanham, MD, Rowman and Littlefield: 327–53.

Chapter 5 is a modified version of earlier work, reproduced here with kind permission from Springer Science+Business Media:

Mallon, R. (2015). "Performance, self-explanation, and agency." *Philosophical Studies* 172: 2777–98. © Springer Science+Business Media Dordrecht 2015.

A portion of Chapter 6, Section 2 derives from:

Mallon, R. (2013b). Naturalistic approaches to social construction. In *Stanford Encyclopedia of Philosophy*. E. N. Zalta. Stanford, CA, Metaphysics Research Lab.

An earlier version of Chapter 8 will be published as:

Mallon, R. (Forthcoming.) "Social construction and achieving reference." *Noûs*. © 2015 Wiley Periodicals, Inc.

Bibliography

Adriaens, P. and A. De Block (2006). "The evolution of a social construction: The case of male homosexuality." *Perspectives in Biology and Medicine* 49: 570–85.

Ahn, W., C. Kalish, S. Gelman et al. (2001). "Why essences are essential in the psychology of concepts." *Cognition* 82: 59–69.

Akins, K. A. and M. E. Windham (1992). "Just science?" *Behavioral and Brain Sciences* 15(2): 376–7.

Alcoff, L. (1997). "Philosophy and racial identity." *Philosophy Today* 41(1): 67–76.

Alcoff, L. (2006). *Visible identities: Race, gender, and the self.* New York; Oxford, Oxford University Press.

American Psychiatric Association (2013). *Diagnostic and statistical manual of mental disorders: DSM-5.* Arlington, VA, American Psychiatric Association.

Amodio, D. M. and P. Devine (2006). "Stereotyping and evaluation in implicit race bias: Evidence for independent constructs and unique effects on behavior." *Journal of Personality and Social Psychology* 91(4): 652–61.

Anderson, E. (2010). *The imperative of integration.* Princeton, NJ, Princeton University Press.

Andreasen, R. O. (1998). "A new perspective on the race debate." *British Journal of the Philosophy of Science* 49: 199–225.

Anscombe, G. E. M. (1957). *Intention.* Ithaca, NY: Cornell University Press.

Appiah, K. A. (1995). The uncompleted argument: Du Bois and the illusion of race. In *Overcoming racism and sexism.* L. A. Bell and D. Blumenfeld. Lanham, MD, Rowman and Littlefield: 59–77.

Appiah, K. A. (1996). Race, culture, identity: Misunderstood connections. In *Color conscious: The political morality of race.* K. A. Appiah and A. Guttmann. Princeton, NJ, Princeton University Press: 192.

Appiah, K. A. (2005). *The ethics of identity.* Princeton, NJ, Princeton University Press.

Aristotle (1947). Politics. In *Introduction to Aristotle.* R. McKeon. New York, Modern Library College Editions: 546–619.

Aspinwall, L. G., T. R. Brown, and J. Tabery (2012). "The double-edged sword: Does biomechanism increase or decrease judges' sentencing of psychopathy?" *Science* 337: 846–9.

Ásta Sveinsdóttir. (2013). "The social construction of human kinds." *Hypatia* 28(4): 716–32.

Astuti, R. (1995). *People of the sea: Identity and descent among the Vezo of Madagascar.* Cambridge; New York, Cambridge University Press.

Astuti, R. (2001). "Are we all natural dualists? A cognitive developmental approach." *Journal of the Royal Anthropological Institute (N.S.)* 7: 429–47.

Astuti, R., G. E. A. Solomon, and S. Carey (2004). "Constraints on cognitive development: A case study of the acquisition of folkbiological and folksociological knowledge in Madagascar." *Monographs of the Society for Research in Child Development* 69(3): i, v, vii–viii, 1–161.

Atran, S. (1990). *Cognitive foundations of natural history*. New York, Cambridge University Press.
Atran, S. (1994). Core domains versus scientific theories: Evidence from systematics and Itza-Maya folkbiology. In *Mapping the mind: Domain specificity in cognition and culture*. L. A. Hirschfeld and S. A. Gelman. New York, Cambridge University Press: 316–40.
Atran, S. (1998). "Folk biology and the anthropology of science: Cognitive universals and cultural particulars." *Behavioral and Brain Sciences* 21: 547–609.
Atran, S. and D. L. Medin (2008). *The native mind and the cultural construction of nature*. Cambridge, MA, MIT Press.
Averill, J. (1980a). A constructivist view of emotion. In *Emotion: Theory, research and experience: Theories of emotion*. R. Plutchik and H. Kellerman. New York, Random House: 1.
Averill, J. (1980b). Emotion and anxiety: Sociocultural, biological, and psychological determinants. In *Explaining emotions*. A. Rorty. Berkeley, CA, University of California Press.
Averill, J. (1994). It's a small world but a large stage. In *The nature of emotion: Fundamental questions*. P. Ekman and R. Davidson. New York, Oxford University Press: 143–5.
Bach, T. (2012). "Gender is a natural kind with a historical essence." *Ethics* 122(2): 231–72.
Banton, M. (1977). *The idea of race*. London, Tavistock Publications.
Banton, M. (1998). *Racial theories*. Cambridge; New York, Cambridge University Press.
Banton, M. and J. Harwood (1975). *The race concept*. New York, Praeger.
Bargh, J. and T. L. Chartrand (1999). "The unbearable automaticity of being." *American Psychologist* 54(7): 462–79.
Bargh, J., M. Chen, and L. Burrows (1996). "Automaticity of social behavior: Direct effects of trait construct and stereotype activation on action." *Journal of Personality and Social Psychology* 71(2): 230–44.
Barrett, H. C. (2005). Adaptations to predators and prey. In *The handbook of evolutionary psychology*. D. M. Buss. New York, Wiley: 200–23.
Barzun, J. (1937). *Race: A study in modern superstition*. New York, Harcourt.
Baumard, N., J.-B. André, and D. Sperber (2013). "A mutualistic approach to morality: The evolution of fairness by partner choice." *Behavioral and Brain Sciences* 36(1): 59–122.
Baumeister, R. F., E. J. Masicampo, and C. N. DeWall (2009). "Prosocial benefits of feeling free: Disbelief in free will increases aggression and reduces helpfulness." *Personality and Social Psychology Bulletin* 35(2): 260–8.
Bayly, S. (1999). *Caste, society, and politics in India*. New York, Cambridge University Press.
Berger, P. L. and T. Luckmann (1966). *The social construction of reality: A treatise in the sociology of knowledge*. Garden City, NY, Doubleday.
Bermúdez, J. L. (2003). "The domain of folk psychology." *Royal Institute of Philosophy Supplement* 53: 25–48.
Bertrand, M. and S. Mullainathan (2004). "Are Emily and Greg more employable than Lakisha and Jamal? A field experiment on labor market discrimination." *American Economic Review* 94(4): 991–1013.

Bird, A. (2002). "Kuhn's wrong turning." *Studies in History and Philosophy of Science Part A* 33(3): 443–63.

Birnbaum, D., I. Deeb, G. Segall, A. Ben-Eliyahu, and G. Diesendruck (2010). "The development of social essentialism: The case of Israeli children's inferences about Jews and Arabs." *Child Development* 81(3): 757–77.

Boehm, C. (2012). *Moral origins: The evolution of virtue, altruism, and shame*. New York, Basic Books.

Bogen, J. (1988). "Symposium papers, comments and an abstract: Comments on 'the sociology of knowledge about child abuse.'" *Noûs* 22(1): 65–6.

Boghossian, P. (2006). *Fear of knowledge: Against relativism and constructivism*. New York, Oxford University Press.

Bloch, M., G. E. A. Solomon, and S. Carey (2001). "Zafimaniry: An understanding of what is passed on from parents to children: A cross-cultural investigation." *Journal of Cognition and Culture* 1(1): 43–68.

Bloom, P. and S. A. Gelman (2008). "Psychological essentialism in selecting the 14th Dalai Lama." *Trends in Cognitive Science* 12(7): 243.

Bloor, D. (1976). *Knowledge and social imagery*. London: Routledge.

Blum, L. (2002). *"I'm not a racist but . . ."* Ithaca, NY, Cornell University Press.

Bourdieu, P. (1990). *The logic of practice*. Stanford, CA, Stanford University Press.

Bowles, S. and H. Gintis (2011). *A cooperative species: human reciprocity and its evolution*. Princeton, NJ, Princeton University Press.

Boyd, R. N. (1983). "On the current status of the issue of scientific realism." *Erkenntnis* 19(1–3): 45–90.

Boyd, R. (1988). How to be a moral realist. In *Essays on moral realism*. G. Sayre-McCord. Ithaca, NY, Cornell University Press: 181–228.

Boyd, R. (1991). Realism, anti-foundationalism and the enthusiasm for natural kinds. *Philosophical Studies* 61: 127–48.

Boyd, R. (1992). Constructivism, realism, and philosophical method. In *Inference, explanation, and other frustrations: Essays in the philosophy of science*. J. Earman. Berkeley, CA, University of California Press: 131–98.

Boyd, R. (1999a). Homeostasis, species and higher taxa. In *Species: New interdisciplinary essays*. Robert A. Wilson. Cambridge, MA, MIT Press: 141–86.

Boyd, R. (1999b). "Kinds, complexity and multiple realization." *Philosophical Studies* 95: 67–98.

Boyer, P. and M. B. Petersen (2011). "The naturalness of (many) social institutions." *Journal of Institutional Economics* 8(1): 1–25.

Brandon, R. N. (1990). *Adaptation and environment*. Princeton, NJ, Princeton University Press.

Brodkin, K. (1998). *How Jews became white folks and what that says about race in America*. New Brunswick, NJ, Rutgers University Press.

Brosig, J. (2002). "Identifying cooperative behavior: Some experimental results in a prisoner's dilemma game." *Journal of Economic Behavior and Organization* 47(3): 275–90.

Bruner, J., L. Postman, and J. Rodrigues (1951). "Expectation and the perception of color." *American Journal of Psychology* LXIV: 216–27.

Buffon, G. L. L., comte de (1791). *Natural history, general and particular*, vol. 7: Chp. 46: Of the degeneration of animals. London, W. Smellie.
Buller, D. J. (2005a). *Adapting minds: Evolutionary psychology and the persistent quest for human nature.* Cambridge, MA, MIT Press.
Buller, D. J. (2005b). "Evolutionary psychology: The emperor's new paradigm." *Trends in Cognitive Sciences* 9(6): 277–83.
Burge, T. (1979). Individualism and the mental. In *Midwest studies in philosophy*, vol. IV. Peter A. French, Theodore E. Uehling, Jr., and Howard K. Wettstein. Minneapolis, MN, University of Minnesota Press: 73–121.
Burge, T. (1986). "Individualism and psychology." *Philosophical Review* XCV(1): 3–45.
Burge, T. (1988). "Individualism and self-knowledge." *Journal of Philosophy* 85: 649–63.
Buss, D. M., R. J. Larsen, D. Westen, and J. Semmelroth (1992). "Sex differences in jealousy: Evolution, physiology, and psychology." *Psychological Science* 3(4): 251–5.
Butler, J. (1990). *Gender trouble: Feminism and subversion of identity.* New York: Routledge.
Carey, S. (1995). On the origins of causal understanding. In *Causal cognition.* D. Sperber, D. Premack, and A. J. Premack. Oxford, Clarendon Press: 268–308.
Carruthers, P. (2006). *The architecture of the mind: Massive modularity and the flexibility of thought.* Oxford, Oxford University Press.
Cartwright, N. (1983). *How the laws of physics lie.* New York, Oxford University Press.
Castleman, M. (2009). "Marital infidelity: How common is it?" *All about Sex* blog, October 15. Retrieved March 6, 2014, from http://www.psychologytoday.com/blog/all-about-sex/200910/marital-infidelity-how-common-is-it.
Cesario, J., J. E. Plaks, and E. T. Higgins (2006). "Automatic social behavior as motivated preparation to interact." *Journal of Personality and Social Psychology* 90(6): 893–910.
Chartrand, T. L. and J. Bargh (1999). "The chameleon effect: The perception-behavior link and social interaction." *Journal of Personality and Social Psychology* 76(6): 893–910.
Chudek, M. and J. Henrich (2011). "Culture-gene coevolution, norm-psychology and the emergence of human prosociality." *Trends in Cognitive Sciences* 15(5): 218–26.
Chudek, M., W. Zhao, and J. Henrich (2013). Culture-gene coevolution, large-scale cooperation, and the shaping of human social psychology. In *Cooperation and its evolution.* K. Sterelny, R. Joyce, B. Calcott, and B. Fraser. Cambridge, MA, MIT Press: 425–58.
Churchland, P. (1981). "Eliminative materialism and the propositional attitudes." *Journal of Philosophy* LXXVII(2): 67–90.
Cikara, M. and J. J. Van Bavel (2014). "The neuroscience of intergroup relations: An integrative review." *Perspectives on Psychological Science* 9: 245–74.
Clark, H. H. (1996). *Using language.* New York, Cambridge University Press.
Cohen, Harold (1949). "An appraisal of the legal tests used to determine who is a negro." *Cornell Law Quarterly* 34: 247.
Collins, H. M. and T. J. Pinch (1993). *The golem: What everyone should know about science.* New York, Cambridge University Press.
Condit, C. M., R. L. Parrott, T. M. Harris, J. Lynch, and T. Dubriwny (2004). "The role of 'genetics' in popular understandings of race in the United States." *Public Understanding of Science* 13: 249–72.

Cooper, R. (2004). "Why Hacking is wrong about human kinds." *British Journal for the Philosophy of Science* 55(1): 73–85.
Cosmides, L. (1989). "The logic of social exchange: Has natural selection shaped how humans reason? Studies with the Wason selection task." *Cognition* 31: 187–276.
Craver, C. F. (2009). "Mechanisms and natural kinds." *Philosophical Psychology* 22(5): 575–94.
Croizet, J. and T. Claire (1998). "Extending the concept of stereotype threat to social class: The intellectual underperformance of students from low socioeconomic backgrounds." *Personality and Social Psychology Bulletin* 24: 588–94.
Daly, M. and M. Wilson (2005). "The 'Cinderella effect' is no fairy tale." *Trends in Cognitive Sciences* 9(11): 507–8.
Davidson, A. (1990). Sex and the emergence of sexuality. In *Forms of desire: Sexual orientation and the social constructionist controversy*. E. Stein. New York, Routledge: 89–132.
Davidson, A. I. (2001). *The emergence of sexuality: Historical epistemology and the formation of concepts*. Cambridge, MA, Harvard University Press.
Davidson, J. N. and M. Smith (1999). "Wittgenstein and Irigaray: Gender and philosophy in a language (game) of difference." *Hypatia* 14(2): 72–96.
Dawkins, R. K. (1978). Animal signals: Information or manipulation? In *Behavioural ecology: An evolutionary approach*. J. R. Krebs and N. B. Davies. Oxford, Blackwell: 282–309.
De Block, A. and B. Du Laing (2007). "Paving the way for an evolutionary social constructivism." *Biological Theory* 2(4): 337–48.
DePaulo, B. M., J. J. Lindsay et al. (2003). "Cues to deception." *Psychological Bulletin* 129(1): 74–118.
Devitt, M. (1981). *Designation*. New York, Columbia University Press.
Devitt, M. (1991). *Realism and truth*. Cambridge, MA: Blackwell.
Devitt, M. (2008). "Resurrecting biological essentialism." *Philosophy of Science* 75(3): 344–82.
Devitt, M. (2010). *Putting metaphysics first: Essays on metaphysics and epistemology*. Oxford; New York, Oxford University Press.
Devitt, M. and K. Sterelny (1999). *Language and reality: An introduction to the philosophy of language*. Cambridge, MA, MIT Press.
Dikötter, Frank (1992). *The discourse of race in modern China*. Stanford, CA, Stanford University Press.
Doris, J. M. (2009). "Skepticism about persons." *Philosophical Issues* 19(1): 57–91.
Doris, J. (2015). *Talking to our selves: Reflection, ignorance, and agency*. Oxford, Oxford University Press.
Doyen, S., O. Klein, C.-L. Pichon, and A. Cleeremans (2012). "Behavioral priming: It's all in the mind, but whose mind?" *PLOS One* 7(1): e29081.
Drabek, M. L. (2014). *Classify and label: The unintended marginalization of social groups*. Lanham, MD, Lexington Books.
Dray, W. H. (1957). *Laws and explanation in history*. London, Oxford University Press.
Dummett, M. (1963/1978). Realism. In *Truth and other enigmas*. Cambridge, MA, Harvard University Press: 145–65.

Dunham, Y., A. S. Baron, and S. Carey (2011). "Consequences of 'minimal' group affiliations in children." *Child Development* 82(3): 793–811.

Dupré, J. (1992). "Blinded by 'science': How not to think about social problems." *Behavioral and Brain Sciences* 15(2): 382–3.

Dupré, J. (2001). *Human nature and the limits of science*. Oxford, Oxford University Press.

Dweck, C. S. (1999). *Self-theories: Their role in motivation, personality, and development*. Philadelphia, PA, Psychology Press.

Egan, A. (2006). "Secondary qualities and self-location." *Philosophy and Phenomenological Research* 72(1): 97–119.

Ekman, P. and W. B. Friesen (1972). Universals and cultural differences in facial expressions of emotion. In *Nebraska symposium on motivation, 1971*. J. K. Cole. Lincoln, NE, University of Nebraska Press: XIX: 207–84.

Ellenberger, H. F. (1970). *The discovery of the unconscious: The history and evolution of dynamic psychiatry*. New York, Basic Books, Harper Collins.

Ereshefsky, M. (2004). "Bridging the gap between human kinds and biological kinds." *Philosophy of Science* 71(5): 912–21.

Ereshefsky, M. (2007). "Foundational issues concerning taxa and taxon names." *Systematic Biology* 56(2): 295–301.

Ereshefsky, M. (2010). "What's wrong with the new biological essentialism." *Philosophy of Science* 77(5): 674–85.

Eshleman, A. S. (2001). "Being is not believing: Fischer and Ravizza on taking responsibility." *Australasian Journal of Philosophy* 79(4): 479–90.

Evans, G. (1973). "The causal theory of names." *Supplementary Proceedings of the Aristotelian Society* 47: 187–208.

Fair, D. (1979). "Causation and the flow of energy." *Erkenntnis* 14(3): 219–50.

Fay, B. (1983). General laws and explaining human behavior. In *Changing social science*. D. R. Sabia and J. Wallulis. Albany, NY, SUNY Press: 103–28.

Feigl, H. (1958). "The 'mental' and the 'physical'." *Minnesota Studies in the Philosophy of Science* 2: 370–497.

Festinger, L., H. W. Riecken, and S. Schachter (1956). *When prophecy fails*. Minneapolis, MN, University of Minnesota Press.

Feyerabend, P. (1975). *Against method*. London: Verso.

Field, H. (1973). "Theory change and the indeterminacy of reference." *Journal of Philosophy* 70: 462–81.

Fine, C. (2010). *Delusions of gender: How our minds, society, and neurosexism create difference*. New York, W. W. Norton.

Fischer, J. M. and M. Ravizza (1998). *Responsibility and control: A theory of moral responsibility*. Cambridge; New York, Cambridge University Press.

Fiske, A. P. (1991). *Structures of social life: The four elementary forms of human relations: Communal sharing, authority ranking, equality matching, market pricing*. New York, Free Press.

Fodor, J. (1981). Special sciences. In *Representations: Philosophical essays on the foundations of cognitive science*. Cambridge, MA, MIT Press: 127–45.

Fodor, J. (1994). *The elm and the expert*. Cambridge, MA, MIT Press.

Fodor, J. A. (1997). "Special sciences: Still autonomous after all these years." *Philosophical Perspectives* 11: 149–63.

Fodor, J. (1998). *Concepts: Where cognitive science went wrong*. New York, Oxford University Press.

Fodor, J. and E. Lepore (1992). *Holism: A shopper's guide*. Cambridge, MA, Blackwell.

Fordham, S. and J. U. Ogbu (1986). "Black students school success: Coping with the burden of acting white." *Urban Review* 18(3): 176–206.

Foucault, M. (1972). The archaeology of knowledge. In *World of man*. New York, Pantheon Books.

Foucault, M. (1978). *The history of sexuality, Vol. I: An introduction*. New York, Pantheon.

Frank, R., T. Gilovich, and D. T. Regan (1993). "The evolution of one-shot cooperation: An experiment." *Ethology and Sociobiology* 14: 247.

Frantz, C. M., A. J. Cuddy, M. Burnett, H. Ray, and A. Hart (2004). "A threat in the computer: The race implicit association test as a stereotype threat experience." *Personality and Social Psychology Bulletin* 30(12): 1611–24.

Fraser, N. and L. Nicholson (1990). Social criticism without philosophy: An encounter between feminism and postmodernism. In *Feminism/postmodernism*. L. Nicholson. New York, Routledge: 19–38.

Fredrickson, G. M. (2002). *Racism: A short history*. Princeton, NJ; Oxford, Princeton University Press.

Fricker, M. (2007). *Epistemic injustice power and the ethics of knowing*. Oxford, Oxford University Press.

Frye, M. (1983). On being white: Toward a feminist understanding of race and race supremacy. In *The politics of reality: Essays in feminist theory*. Freedom, CA, Crossing Press: 110–28.

Fryer, R. and P. Torelli (2010). "An empirical analysis of acting white." *Journal of Public Economics* 94(5–6): 380–96.

Gannett, L. (2001). "Racism and human genome diversity research: The ethical limits of 'population thinking.'" *Philosophy of Science* 68(Proceedings): S479–92.

Gannett, L. (2010). "Questions asked and unasked: How by worrying less about the 'really real' philosophers of science might better contribute to debates about genetics and race." *Synthese* 177: 363–85.

Geertz, C. (1973). Thick description: Toward an interpretive theory of culture. In *The interpretation of cultures: Selected essays*. New York, Basic Books: 3–32.

Gelman, S. A. (2003). *The essential child: Origins of essentialism in everyday thought*. New York, Oxford University Press.

Gelman, S. A. and J. D. Coley (1990). "The importance of knowing a dodo is a bird: Categories and inferences in 2-year-old children." *Developmental Psychology* 26: 796–804.

Gelman, S., J. D. Coley, and G. M. Gottfried (1994). Essentialist beliefs in children: The acquisition of concepts and theories. In *Mapping the mind*. L. A. Hirschfeld and S. Gelman. New York, Cambridge University Press: 341–66.

Gelman, S. and L. Hirschfeld (1999). How biological is essentialism? In *Folkbiology*. S. Atram and D. Medin. Cambridge, MA, MIT Press.

Gelman, S. A. and E. Markman (1986). "Categories and induction in young children." *Cognition* 23: 183–209.

Gelman, S. A. and H. M. Wellman (1991). "Insides and essences: Early understandings of the non-obvious." *Cognition* 38: 213–44.

Gendler, T. (2008). "Alief and belief." *Journal of Philosophy* 105(10): 634–63.

Ghiselin, M. T. (1974). "A radical solution to the species problem." *Systematic Zoology* 23: 536–44.

Gibson, C. E., J. Losee, and C. Vitiello (2014). "A replication attempt of stereotype susceptibility (Shih, Pittinsky, and Ambady, 1999): Identity salience and shifts in quantitative performance." *Social Psychology* 45(3): 194–8.

Gil-White, F. (2001a). "Are ethnic groups biological 'species' to the human brain?" *Current Anthropology* 42(4): 515–54.

Gil-White, F. (2001b). "Sorting is not categorization: A critique of the claim that Brazilians have fuzzy racial categories." *Cognition and Culture* 1(3): 219–49.

Gilbert, M. (1989). *On social facts*. Princeton, NJ, Princeton University Press.

Giner-Sorolla, R., J. Embley, and L. Johnson (2015). "Replication of Vohs and Schooler (2008, PS, Study 1)." Open Science Framework. July 24. osf.io/i29mh.

Ginet, C. (2006). "Working with Fischer and Ravizza's account of moral responsibility." *Journal of Ethics* 10: 229–53.

Glasgow, J. (2009). *A theory of race*. New York, Routledge.

Gobineau, A. (1999). *The inequality of human races*. New York, H. Fertig.

Godfrey-Smith, P. (2009). *Darwinian populations and natural selection*. New York, Oxford University Press.

Goffman, E. (1970). *Asylums: Essays on the social situation of mental patients and other inmates*. Chicago, IL, Aldine.

Goldberg, D. T. (1993). *Racist culture: Philosophy and the politics of meaning*. Cambridge, MA, Blackwell.

Gonzales, P. M., H. Blanton, and K. J. Williams (2002). "The effects of stereotype threat and double-minority status on the test performance of Latino women." *Personality and Social Psychology Bulletin* 28: 659–70.

Gooding-Williams, R. (1988). "Race, multiculturalism and democracy." *Constellations* 5: 18–41.

Goodman, N. (1978). *Ways of worldmaking*. Indianapolis, IN, Hackett Publishing.

Gossett, T. F. (1963). *Race: The history of an idea in America*. Dallas, TX, Southern Methodist University Press.

Gottfried, G. M. and S. A. Gelman (2005). "Developing domain-specific causal-explanatory frameworks: The role of insides and immanence." *Cognitive Development* 20: 137–58.

Greene, J. D. (2013). *Moral tribes: Emotion, reason, and the gap between us and them*. New York, Penguin Press.

Greene, J. D., R. B. Sommerville, L. E. Nystrom, J. M. Darley, and J. D. Cohen (2001). "An fMRI investigation of emotional engagement in moral judgment." *Science* 293: 2105–8.

Greenwald, A. G., M. R. Banaji, and B. A. Nosek (2015). "Statistically small effects of the implicit association test can have societally large effects." *Journal of Personality and Social Psychology* 108(4): 553–61.

Greenwald, A. G., D. E. McGhee, and J. L. K. Schwartz (1998). "Measuring individual differences in implicit cognition: The implicit association test." *Journal of Personality and Social Psychology* 74: 1464–80.

Griffith, A. (ms). "The metaphysics of social construction: A 'grounding' account."

Griffiths, P. E. (1997). *What Emotions Really Are*. Chicago, IL, University of Chicago Press.

Griffiths, P. E. (1999). Squaring the circle: Natural kinds with historical essences. In *Species: New interdisciplinary essays*. R. A. Wilson. Cambridge, MA, MIT Press: 209–28.

Griffiths, P. E. (2002). "What is innateness." *Monist* 85(1): 70–85.

Guillaumin, C. (1995). *Racism, sexism, power, and ideology*. London; New York, Routledge.

Hacking, I. (1986). Making up people. In *Reconstructing individualism: Autonomy, individuality, and the self in Western thought*. T. C. Heller, M. Sosna, and D. E. Wellbery. Stanford, CA, Stanford University Press: 222–36.

Hacking, I. (1991). "The making and molding of child abuse." *Critical Inquiry* 17: 253–88.

Hacking, I. (1992a). "Multiple personality disorder and its hosts." *History of the Human Sciences* 5(2): 3–31.

Hacking, I. (1992b). World making by kind making: Child-abuse for example. *In How classification works: Nelson Goodman among the social sciences*. M. Douglas and D. Hull. Edinburgh, Edinburgh University Press: 180–238.

Hacking, I. (1995a). *Rewriting the soul: Multiple personality and the sciences of memory*. Princeton, NJ, Princeton University Press.

Hacking, I. (1995b). The looping effects of human kinds. In *Causal cognition: A multidisciplinary debate*. D. Sperber, D. Premack, and A. J. Premack. New York, Clarendon Press: 351–94.

Hacking, I. (1998). *Mad travelers: Reflections on the reality of transient mental illnesses*. Charlottesville, VA, University Press of Virginia.

Hacking, I. (1999). *The social construction of what?* Cambridge, MA, Harvard University Press.

Hacking, I. (2005). "Why race still matters." *Daedelus* 134: 102–16.

Hacking, I. (2007). "Kinds of people: Moving targets." *Proceedings of the British Academy* 151: 285–318.

Haidt, J. (2001). "The emotional dog and its rational tail: A social intuitionist approach to moral judgment." *Psychological Review* 108: 814–34.

Hall, S. (1996). New ethnicities. In *Stuart Hall: Critical dialogues in cultural studies*. D. Morley and K.-H. Chen. London; New York, Routledge: 442–51.

Hannaford, I. (1996). *Race: The history of an idea in the West*. Washington, DC, Woodrow Wilson Center Press.

Hardimon, M. (2003). "The ordinary concept of race." *Journal of Philosophy* C(9): 437–55.

Harding, S. (2015). "After Mr. Nowhere: What kind of proper self for a scientist?" *Feminist Philosophy Quarterly* 1(2).

Harman, G. (1977). "Review of linguistic behavior by Jonathan Bennett." *Language* 53: 417–24.

Harris, A. L. (2011). *Kids don't want to fail: Oppositional culture and the black-white achievement gap.* Cambridge, MA, Harvard University Press.

Harris, M. (1970). "Referential ambiguity in the calculus of Brazilian racial identity." *Southwestern Journal of Anthropology* 26(1): 1–14.

Harris, M. and C. Kottack (1963). "The structural significance of Brazilian categories." *Sociologia* 25: 203–8.

Haslam, N., B. Bastian, P. Bain, and Y. Kashima (2006). "Psychological essentialism, implicit theories, and intergroup relations." *Group Processes and Intergroup Relations* 9(1): 63–76.

Haslam, N., L. Rothschild, and D. Ernst (2000). "Essentialist beliefs about social categories." *British Journal of Social Psychology* 39: 113–27.

Haslanger, S. (1995). "Ontology and social construction." *Philosophical Topics* 23(2): 95–125. Reprinted in Haslanger (2012).

Haslanger, S. (2000). "Gender and race: (What) are they? (What) do we want them to be?" *Noûs* 34: 31–55. Reprinted in Haslanger (2012).

Haslanger, S. (2003). Social construction: The "debunking" project. In *Socializing metaphysics: The nature of social reality.* F. Schmitt. Lanham, MD, Rowman and Littlefield: 301–25. Reprinted in Haslanger (2012).

Haslanger, S. (2005). "What are we talking about? The semantics and politics of social kinds." *Hypatia* 20(4): 10–26. Reprinted in Haslanger (2012).

Haslanger, S. A. (2012). *Resisting reality: Social construction and social critique.* New York, Oxford University Press.

Heidegger, M. (2010). *Being and time.* Albany, NY, State University of New York Press.

Henrich, J., S. J. Heine, and A. Norenzayan (2010). "The weirdest people in the world." *Behavioral and Brain Sciences* 33: 61–135.

Henrich, J. and R. McElreath (2003). "The evolution of cultural evolution." *Evolutionary Anthropology* 12: 123–35.

Hieronymi, P. (2005). "The wrong kind of reason." *Journal of Philosophy* 102(9): 437–57.

Hirschfeld, L. A. (1995). "Do children have a theory of race?" *Cognition* 54: 209–52.

Hirschfeld, L. A. (1996). *Race in the making: Cognition, culture, and the child's construction of human kinds.* Cambridge, MA, MIT Press.

Hoff, K., M. Kshetramade, and E. Fehr (2011). "Caste and punishment: The legacy of caste culture in norm enforcement." *Economic Journal* 121: F449–75.

Hoyt, C. L., J. L. Burnette, and L. Auster-Gussman (2014). "'Obesity is a disease': Examining the self-regulatory impact of this public-health message." *Psychological Science* 25(4): 997–1002.

Hull, D. (1976). "Are species really individuals?" *Systematic Zoology* 25: 174–91.

Hull, D. (1978). "A matter of individuality." *Philosophy of Science* 45: 335–60.

Inatiev, N. (1995). *How the Irish became white.* New York, Routledge.

Isaac, B. H. (2004). *The invention of racism in classical antiquity.* Princeton, NJ, Princeton University Press.

Isaacson, W. and W. Allen (2001). The heart wants what it wants. *Time*, August 31.

Jackson, F. (1998). *From metaphysics to ethics: A defense of conceptual analysis.* Oxford, Oxford University Press.

Jackson, F. and P. Pettit (1988). "Functionalism and broad content." *Mind* 97: 381–400.

Jayaratne, T. E., S. A. Gelman, M. Feldbaum, J. P. Sheldon, E. M. Petty, and S. L. R. Kardia (2009). "The perennial debate: Nature, nurture, or choice? Black and white Americans' explanations for individual differences." *Review of General Psychology* 13(1): 24–33.

Jayaratne, T. E., O. Ybarra, J. P. Sheldon, T. N. Brown, M. Feldbaum, C. A. Pfeffer, and E. M. Petty (2006). "White Americans' genetic lay theories of race differences and sexual orientation: Their relationship with prejudice toward blacks, and gay men and lesbians." *Group Processes and Intergroup Relations* 9(1): 77–94.

Jenkins, C. S. (2005). "Realism and independence." *American Philosophical Quarterly* 42(3): 199–209.

Johnson, S. C. and G. E. A. Solomon (1997). "Why dogs have puppies and cats have kittens: The Role of birth in young children's understanding of biological origins." *Child Development* 68(3): 404–19.

Johnston, A. I. (1995). *Cultural realism: Strategic culture and grand strategy in Chinese history*. Princeton, NJ, Princeton University Press.

Jones, D. (2009). "Looks and living kinds: Varieties of racial cognition in Bahia, Brazil." *Journal of Cognition and Culture* 9: 247–69.

Kahneman, D. (2011). *Thinking, fast and slow*. New York, Farrar, Straus and Giroux.

Kamtekar, R. (2002). Distinction without a difference? Plato on "genos" vs. "race." In *Philosophers on race: Critical essays*. J. K. Ward and T. Lott. New York, Blackwell: 1–13.

Kanovsky, M. (2007). "Essentialism and folksociology: Ethnicity again." *Journal of Cognition and Culture* 7: 241–81.

Kaplan, J. M. (2010). "When socially determined categories make biological realities." *Monist* 93(2): 283–99.

Keil, F. C. (1989). *Concepts, kinds, and cognitive development*. Cambridge, MA, MIT Press, Bradford Books.

Keil, F. (1992). The origins of an autonomous biology. In *Modularity and constraints in language and cognition: The Minnesota symposia*. M. Gunnar and M. Maratsos. Hilldale, NJ, Earlbaum: 103–37.

Kelly, D. (2011). *Yuck! The nature and moral significance of disgust*. Cambridge, MA, MIT Press.

Kelly, D., L. Faucher, and E. Machery (2010). "Getting rid of racism: Assessing three proposals in light of psychological evidence." *Journal of Social Philosophy* 41(3): 293–322.

Khalidi, M. A. (2013). *Natural categories and human kinds: Classification in the natural and social sciences*. New York, Cambridge University Press.

Kim, J. (1998). *Mind in a physical world: An essay on the mind-body problem and mental causation*. Cambridge, MA, MIT Press.

Kinzler, K. D., K. Shutts, J. DeJesus, and E. S. Spelke (2009). "Accent trumps race in guiding children's social preferences." *Social Cognition* 27: 623–34.

Kitcher, P. (1978). "Theories, theorists and theoretical change." *Philosophical Review* 87(4): 519–47.

Kitcher, P. (1993). *The advancement of science: Science without legend, objectivity without illusions*. New York, Oxford University Press.

Kitcher, P. (1999). Race, ethnicity, biology, culture. In *Racism*. L. Harris. New York, Humanity Books: 87–120.

Kitcher, P. (2011). *The ethical project*. Cambridge, MA, Harvard University Press.
Klein, P. D. (1999). "Human knowledge and the infinite regress of reasons." *Philosophical Perspectives* 13(s13): 297–325.
Kornblith, H. (1993). *Inductive inference and its natural ground: An essay in naturalistic epistemology*. Cambridge, MA, MIT Press.
Kripke, S. (1972/1980). *Naming and necessity*. Cambridge, MA, Harvard University Press.
Kuhn, T. S. (1962/1970). *The structure of scientific revolutions*. Chicago, IL, University of Chicago Press.
Kuhn, T. S. (1977). Objectivity, value judgment, and theory choice. In *The essential tension*, Chicago, IL, University of Chicago Press: 320–39.
Kuhn, T. S., J. Conant, and J. Haugeland (2000). *The road since structure: Philosophical essays, 1970-1993, with an autobiographical interview*. Chicago, IL; London, University of Chicago Press.
Kukla, A. (2000). *Social constructivism and the philosophy of science*. London, Routledge.
Kunda, Z. (1999). *Social cognition: Making sense of people*. Cambridge, MA, MIT Press.
Kuorikoski, J. and S. Pöyhönen (2012). "Looping kinds and social mechanisms." *Sociological Theory* 30(3): 187–205.
Kurth, C. (2015). "Anxiety, normative uncertainty, and social regulation." *Biology and Philosophy*, online only.
Kurzban, R., J. Tooby, and L. Cosmides (2001). "Can race be erased? Coalitional computation and social categorization." *Proceeding of the National Academy of Science* 98(26): 15387–92.
Laland, K. N., J. Odling-Smee, and M. W. Feldman (2000). "Niche construction, biological evolution and cultural change." *Behavioral and Brain Sciences* 23: 131–46.
Laland, K. N. and K. Sterelny (2006). "Seven reasons (not) to neglect niche construction." *Evolution* 60(9): 1751–62.
Laqueur, T. W. (1990). *Making sex: Body and gender from the Greeks to Freud*. Cambridge, MA, Harvard University Press.
Latour, B. and S. Woolgar (1986). *Laboratory life: The construction of scientific facts*. Princeton, NJ, Princeton University Press.
Laudan, L. (1984). *Science and values: The aims of science and their role in scientific debate*. Berkeley, CA, University of California Press.
Lazarus, R. (1994). Universal antecedents of the emotions. In *The nature of emotion: Fundamental questions*. P. Ekman and R. Davidson. New York, Oxford University Press: 163–71.
Leslie, A. M. (1994). ToMM, ToBY, and Agency: Core architecture and domain specificity. In *Mapping the mind: Domain specificity in cognition and culture*. L. A. Hirschfeld and S. A. Gelman. New York, Cambridge University Press: 119–48.
Leslie, S.-J. (forthcoming). "The original sin of cognition: Race, prejudice and generalization." *Journal of Philosophy*.
Levy, A. (2009). Nora Ephron knows what to do. *New Yorker*, July 6.
Levy, S. R., S. J. Stroessner, and C. S. Dweck (1998). "Stereotype formation and endorsement: The role of implicit theories." *Journal of Personality and Social Psychology* 74(6): 1421–36.
Lewis, D. (1969). *Convention: A philosophical study*. Newcastle, Basil Blackwell.

Lewis, D. (1970). "How to define theoretical terms." *Journal of Philosophy* 67: 426–46.
Lewis, D. (1972). "Psychophysical and theoretical identifications." *Australasian Journal of Philosophy* 50: 249–58.
Lewis, D. (1973). "Causation." *Journal of Philosophy* 70(17): 556–67.
Lewis, D. (1978). "Truth in fiction." *American Philosophical Quarterly* 15(1): 37–46.
Lewis, D. (2000). "Causation as influence." *Journal of Philosophy* 97(4): 182–97.
Lewontin, R. (1982). *Human diversity*. New York: Scientific American Press.
Longino, H. E. (1990). Science as social knowledge: Values and objectivity in scientific inquiry. Princeton, NJ, Princeton University Press.
Lukács, G. (1923/1971). Reification and the consciousness of the proletariat. In *History and class consciousness*. R. Livingstone (trans.). Cambridge, MA, MIT Press: 83–222.
Lutz, C. (1988). Unnatural emotions: Everyday sentiments on a Micronesian atoll and their challenge to Western thoery. Chicago, IL, University of Chicago Press.
Lycan, W. (1988). *Judgement and justification*. Cambridge, Cambridge University Press.
Machery, E. (2008). "A plea for human nature." *Philosophical Psychology* 21: 321–30.
Machery, E. (2014). Social ontology and the objection from reification. In *Perspectives on social ontology and social cognition*. M. Gallotti and J. Michael. Dordrecht, Springer: 87–102.
Machery, E. and L. Faucher (2005a). "Social construction and the concept of race." *Philosophy of Science* 72: 1208–19.
Machery, E. and L. Faucher (2005b). Why do we think racially? In *Handbook of categorization in cognitive science*. H. Cohen and C. Lefebvre. Orlando, FL, Elsevier: 1010–33.
Machery, E., L. Faucher, and D. Kelly (2010). "On the alleged inadequacies of psychological explanations of racism." *Monist* 93(2): 228–54.
Machery, E., R. Mallon, S. Nichols, and S. Stich (2004). "Semantics, cross-cultural style." *Cognition* 92: B1–12.
Machery, E., C. Y. Olivola, H. Cheon, I. T. Kurniawan, C. Mauro, N. Struchiner, and H. Susianto (ms). "Is folk essentialism a fundamental feature of human cognition?"
MacKinnon, C. A. (1987). *Feminism unmodified: Discourses on life and law*. Cambridge, MA, Harvard University Press.
MacKinnon, C. A. (1989). *Toward a feminist theory of the state*. Cambridge, MA, Harvard University Press.
Mahalingam, R. and J. Rodriguez (2006). "Culture, brain transplants and implicit theories of identity." *Journal of Cognition and Culture* 6(3–4): 452–62.
Maibom, H. L. (2007). "Social systems." *Philosophical Psychology* 20: 1–22.
Malle, B. F. (2006). "The actor-observer asymmetry in causal attribution: A (surprising) meta-analysis." *Psychological Bulletin* 132: 895–919.
Mallon, R. (2003). Social construction, social roles and stability. In *Socializing metaphysics*. F. Schmitt. Lanham, MD, Rowman and Littlefield: 327–53.
Mallon, R. (2004). "Passing, traveling, and reality: Social construction and the metaphysics of race." *Noûs* 38(4): 644–73.
Mallon, R. (2006). "'Race': Normative, not metaphysical or semantic." *Ethics* 116(3): 525–51.
Mallon, R. (2007a). "Human categories beyond non-essentialism." *Journal of Political Philosophy* 15(2): 146–68.

Mallon, R. (2007b). "Arguments from reference and the worry about dependence." *Midwest Studies in Philosophy* XXXI: 160–83.

Mallon, R. (2007c). "A field guide to social construction." *Philosophy Compass* 2(1): 93–108.

Mallon, R. (2009). "Commentary on Joshua Glasgow's *A Theory of Race*." *Symposia on Gender, Race and Philosophy* 5(2).

Mallon, R. (2010). "Sources of racialism." *Journal of Social Philosophy* 41(3): 272–92.

Mallon, R. (2013a). "Was *Race* thinking invented in the modern West?" *Studies in History and Philosophy of Science* 44: 77–88.

Mallon, R. (2013b). Naturalistic approaches to social construction. In *Stanford encyclopedia of philosophy*. E. N. Zalta. Stanford, CA, Metaphysics Research Lab.

Mallon, R. (forthcoming). "Social Roles and Reification." In Routledge Handbook of Social Mind. J. Kiverstein. New York.

Mallon, R. and D. Kelly (2012). Making race out of nothing: Psychologically constrained social roles. In *Oxford handbook of philosophy of social science*. H. Kincaid. Oxford, Oxford University Press: 507–32.

Mallon, R., E. Machery, S. Nichols, and S. Stich (2009). "Against arguments from reference." *Philosophy and Phenomenological Research* 79(2): 332–56.

Mallon, R. and S. P. Stich (2000). "The odd couple: The compatibility of social construction and evolutionary psychology." *Philosophy of Science* 67: 133–54.

Mallon, R. and J. Weinberg (2006). "Innateness as closed process invariance." *Philosophy of Science* 73: 323–44.

Mann, S. and A. Vrij (2006). "Police officers' judgements of veracity, tenseness, cognitive load and attempted behavioural control in real life police interviews." *Psychology, Crime, and Law* 12: 307–19.

Marmot, M. G., G. Rose, M. Shipley, and P. J. Hamilton (1978). "Employment grade and coronary heart disease in British civil servants." *Journal of Epidemiology and Community Health* 32(4): 244–9.

Marston, C. and R. Lewis (2014). "Anal heterosex among young people and implications for health promotion: A qualitative study in the UK." *BMJ Open* 4(8): 1–6.

Martin, C. L. and S. Parker (1995). "Folk theories about sex and race differences." *Personality and Social Psychology Bulletin* 21: 45–57.

Martin, J. R. (1994). "Methodological essentialism, false difference, and other dangerous traps." *Signs* 19: 630–57.

Marx, K. (1904). *A Contribution to the Critique of Political Economy*. Trans. N.I. Stone. Chicago, Charles Kerr and Company. International Library Publishing Company.

Mayr, E. (1976). Typological versus population thinking. In *Evolution and the diversity of life*. Cambridge, MA, Harvard University Press: 26–9.

Mayr, E. (1984). Species concepts and their application. In *Conceptual issues in evolutionary biology: An anthology*. E. Sober. Cambridge, MA, MIT Press: 531–40.

McKenna, W. (1984). "Review: The construction of desire." *Women's Review of Books* 1(6): 3–5.

McKenna, M. (2000). "Assessing reasons-responsive compatibilism: J. M. Fischer and M. Ravizza's responsibility and control." *International Journal of Philosophical Studies* 8(1): 89–114.

Mercier, H. and D. Sperber (2011). "Argumentation: Its adaptiveness and efficacy." *Behavioral and Brain Sciences* 34(2): 94–111.

Miller, A. (2014). Realism. In *Stanford encyclopedia of philosophy*. E. N. Zalta. Stanford, CA, Metaphysics Research Lab.

Millikan, R. G. (1999). "Historical kinds and the 'special sciences.'" *Philosophical Studies* 95(1–2): 45–65.

Mills, C. (1998). *Blackness visible: Essays on philosophy and race.* Ithaca, NY, Cornell University Press.

Monahan, J. and G. L. Hood (1976). "Psychologically disordered and criminal offenders: Perceptions of their volition and responsibility." *Criminal Justice and Behavior* 3(2): 123–34.

Monterosso, J., E. B. Royzman, and B. Schwartz (2005). "Explaining away responsibility: Effects of scientific explanation on perceived culpability." *Ethics and Behavior* 15(2): 139–58.

Moon, A. and S. S. Roeder (2014). "A secondary replication attempt of stereotype susceptibility (Shih, Pittinsky, and Ambady, 1999)." *Social Psychology* 45(3): 199–201.

Moss-Racusin, C. A., J. F. Dovidio, V. L. Brescoll, M. J. Graham, and J. Handelsman (2012). "Science faculty's subtle gender biases favor male students." *PNAS Proceedings of the National Academy of Sciences of the United States of America* 109(41): 16474–9.

Mottola, G. (2009). *Adventureland*. Film.

Moya, C. and R. Boyd (2015). "Different selection pressures give rise to distinct ethnic phenomena: A functionalist framework with illustrations from the Peruvian Altiplano." *Human Nature* 26: 1–27.

Murphy, D. (2006). *Psychiatry in the scientific image*. Cambridge, MA, MIT Press.

Murray, D. and E. Nahmias (2014). "Explaining away incompatibilist intuitions." *Philosophy and Phenomenological Research* 88(2): 434–67.

Nahmias, E. (2007). Autonomous agency and the threat of social psychology. In *Cartographies of the mind: Philosophy and psychology in intersection*. M. Marraffa, M. Caro, and F. Ferretti. New York, Springer.

Nahmias, E. A., T. Kvaran, and D. J. Coates (2007). "Free will, moral responsibility, and mechanism: Experiments on folk intuitions." *Midwest Studies in Philosophy* XXXI: 214–42.

Nichols, S. (2014). "Process debunking and ethics." *Ethics* 124(4): 727–49.

Nichols, S., À. Pinillos, and R. Mallon (forthcoming). "Referential ambiguity." *Mind*.

Nichols, S. and S. P. Stich (2003). *Mindreading: An integrated account of pretence, self-awareness, and understanding other minds*. Oxford: Oxford University Press.

Nicholson, L. (1994). "Interpreting gender." *Signs* 20(1): 79–105.

Nicholson, L. (1999). *The play of reason: From the modern to the postmodern*. Ithaca, NY: Cornell University Press.

Nisbett, R. E. and N. Bellows (1977). "Verbal reports about causal influences on social judgments: Private access vs. public theories." *Journal of Personality and Social Psychology* 35: 613–24.

Nisbett, R. E. and T. D. Wilson (1977). "Telling more than we can know: Verbal reports on mental processes." *Psychological Review* 84(3): 231–59.

Nogueira, O. (1955). "Preconceito racial de marca e preconceito racial de origem." *Anais do XXXI Congresso Internacional de Americanistas* 1.

Nolan, J. M., P. W. Schultz, R. B. Cialdini, N. J. Goldstein, and V. Griskevicius (2008). "Normative social influence is underdetected." *Personality and Social Psychology Bulletin* 34(7): 913–23.

Nosek, B., M. R. Banaji, and A. G. Greenwald (2002a). "Harvesting implicit group attitudes and beliefs from a demonstration website." *Group Dynamics* 6(1): 101–15.

Nosek, B., M. R. Banaji, and A. G. Greenwald (2002b). "Math = male, me = female, therefore math ≠ me." *Journal of Personality and Social Psychology* 83(1): 44–59.

Nosek, B. A., F. L. Smyth, J. J. Hansen et al. (2007). "Pervasiveness and correlates of implicit attitudes and stereotypes." *European Review of Social Psychology* 18: 36–88.

Odling-Smee, F. J., K. N. Laland, and M. W. Feldman (2003). *Niche construction: The neglected process in evolution*. Princeton, NJ, Princeton University Press.

Olivola, C. Y. and E. Machery (2014). "Is psychological essentialism an inherent feature of human cognition?" *Behavioral and Brain Sciences* 37(5): 499.

Omi, M. and H. Winant. (1986). *Racial formation in the United States: From the 1960s to the 1980s*. New York, Routledge and Kegan Paul.

Omi, M. and H. Winant. (1994). *Racial formation in the United States: From the 1960s to the 1990s*. New York, Routledge.

Omi, M. and H. Winant (2014). *Racial formation in the United States*. New York, Routledge.

Oswald, F. L., G. Mitchell, H. Blanton, J. Jaccard, and P. E. Tetlock (2013). "Predicting ethnic and racial discrimination: A meta-analysis of IAT criterion studies." *Journal of Personality and Social Psychology* 105: 171–92.

Outlaw, L. (1996). *On race and philosophy*. New York, Routledge.

Papineau, D. (2010). "Realism, Ramsey sentences and the pessimistic meta-induction." *Studies in History and Philosophy of Science Part A* 41(4): 375–85.

Pashler, H., C. Harris, and N. Coburn (2011). "Elderly-related words prime slow walking." Retrieved August 8, 2013, from <http://www.PsychFileDrawer.org/replication.php?attempt=MTU%3D>.

Paul, L. A. and N. Hall (2013). *Causation: A user's guide*. Oxford, Oxford University Press.

Payne, B. K. (2001). "Prejudice and perception: The role of automatic and controlled processes in misperceiving a weapon." *Journal of Personality and Social Psychology* 81: 181–92.

Peterson, R. T. (2010). *Peterson field guide to birds of Eastern and Central North America*. Boston, MA; New York, Houghton Mifflin Harcourt.

Pettit, P. (1991). "Realism and response-dependence." *Mind* 100(4): 587–626.

Pettit, P. (2003). Groups with minds of their own. In *Socializing metaphysics: The nature of social reality*. F. Schmitt. Lanham, MD, Rowman and Littlefield: 167–93.

Pickering, A. (1984). *Constructing quarks: A sociological history of particle physics*. Edinburgh, Edinburgh University Press.

Pietraszewski, D., L. Cosmides, and J. Tooby (2014). "The content of our cooperation, not the color of our skin: An alliance detection system regulates categorization by coalition and race, but not sex." *PLoS ONE* 9(2): e88534.

Pinker, S. (2002). *The blank slate: The modern denial of human nature*. New York, Viking.

Piper, A. (1992). "Passing for white, passing for black." *Transition* 58: 4–32.

Plato (1961). Cratylus. Trans. B. Jowett. In *The collected dialogues of Plato including the letters*. E. Hamilton and H. Cairns. Princeton, NJ, Princeton University Press: 421–74.

Price, J. and J. Wolfers (2010). "Racial discrimination among NBA referees." *Quarterly Journal of Economics* 125(4): 1859–87.

Prinz, J. (2007). *The emotional construction of morals*. New York, Oxford University Press.

Putnam, F. W. (1989). *Diagnosis and treatment of multiple personality disorder*. New York, Guilford Press.

Putnam, H. (1975). The meaning of "meaning." In *Mind, language and reality: Philosophical papers*. New York, Cambridge University Press: 215–71.

Putnam, H. (1982). "Why there isn't a ready-made world." *Synthese* 51(2): 141–67.

Putnam, H. (1990). *Realism with a human face*. Cambridge, MA, Harvard University Press.

Puzzo, D. A. (1964). "Racism and the Western tradition." *Journal of the History of Ideas* 25(4): 579–86.

Quine, W. V. O. (1953). Two dogmas of empiricism. In *From a logical point of view*. Cambridge, MA, Harvard University Press: 20–46.

Rhodes, M., S. J. Leslie, and C. M. Tworek (2012). "Cultural transmission of social essentialism." *Proceedings of the National Academy of Sciences* 109(34): 13526–31.

Richerson, P. J. and R. Boyd (2005). *Not by genes alone how culture transformed human evolution*. Chicago, IL, University of Chicago Press.

Richerson, P. J. and J. Henrich (2011). "Tribal social instincts and the cultural evolution of institutions to solve collective action problems." *Cliodynamics: Journal of Theoretical and Mathematical History* 3(1): 38–80.

Robbins, P. (2006). "The ins and outs of introspection." *Philosophy Compass* 1(6): 617–30.

Robinson, E. (1999). *Coal to cream: A black man's journey beyond color to an affirmation of race*. New York, Free Press.

Root, M. (2000). "How we divide the world." *Philosophy of Science* 67 (Proceedings): 628–39.

Rorty, R. (1979). *Philosophy and the mirror of nature*. Princeton, NJ, Princeton University Press.

Rorty, R. (1998). *Truth and progress*. Cambridge; New York, Cambridge University Press.

Rosen, G. (1994). Objectivity and modern idealism: What is the question? In *Philosophy in mind: The place of philosophy in the study of mind*. M. Michael and J. O'Leary-Hawthorne. Boston, MA, Kluwer Academic Publishers: 277–319.

Rothbart, M. and M. Taylor (1992). Category labels and social reality: Do we view social categories as natural kinds? In *Language, interaction and social cognition*. G. R. Semin and K. Fiedler. London, Sage Publications: 11–36.

Salmon, W. C. (1994). "Causality without counterfactuals." *Philosophy of Science* 61(2): 297–312.

Sartre, J.-P. (1956). *Being and nothingness: An essay on phenomenological ontology*. New York, Philosophical Library.

Sayre-McCord, G. (1988). *Essays on moral realism*. Ithaca, NY, Cornell University Press.

Schaffer, J. (2009). On what grounds what. In *Metametaphysics: New essays on the foundations of ontology*. D. Manley, D. J. Chalmers, and R. Wasserman. Oxford, Oxford University Press: 347–83.

Scheff, T. J. (1974). "The labelling theory of mental illness." *American Sociological Review* 39: 444–52.

Scheff, T. (1984). *Being mentally ill: A sociological theory*. New York, Aldine Publishing.

Schiffer, S. (1972). *Meaning*. Oxford, Oxford University Press.

Schmader, T., M. Johns, and C. Forbes (2008). "An integrated process model of stereotype threat effects on performance." *Psychological Review* 115: 336–56.

Schneider, D. J. (2004). *The psychology of stereotyping*. New York; London, Guilford Press.

Schroeter, L. (2004). "The limits of conceptual analysis." *Pacific Philosophical Quarterly* 85(4): 425–53.

Searle, J. R. (1964). "How to derive 'ought' from 'is.'" *Philosophical Review* 73(1): 43–58.

Searle, J. (1990). Collective intentions and actions. In *Intentions in communication*. P. R. C. J. Morgan and M. Pollack. Cambridge, MA, MIT Press: 401–15.

Searle, J. (1995). *The construction of social reality*. New York, Free Press.

Searle, J. R. (2010). *Making the social world: The structure of human civilization*. Oxford; New York, Oxford University Press.

Sesardic, N. (2003). "Book review of N. Zack, Philosophy of Science and Race." *Philosophy of Science* 70: 447–9.

Sesardic, N. (2010). "Race: A social destruction of a biological concept." *Biology and Philosophy* 25: 143–62.

Shackelford, T. K. and D. M. Buss (1996). "Betrayal in mateships, friendships, and coalitions." *Personality and Social Psychology Bulletin* 22(11): 1151–64.

Shafer-Landau, R. (2003). *Moral realism: A defence*. Oxford, Oxford University Press.

Shariff, A. F., J. D. Greene, J. C. Karremans et al. (2014). "Free will and punishment: A mechanistic view of human nature reduces retribution." *Psychological Science* 25(8): 1563–70.

Shelby, T. (2002). "Foundations of black solidarity: Collective identity or common oppression?" *Ethics* 112: 231–66.

Shelby, T. (2005). *We who are dark: The philosophical foundations of Black solidarity*. Cambridge, MA, Belknap Press of Harvard University Press.

Sheldon, J. P., T. E. Jayaratne, and E. M. Petty (2007). "White Americans' genetic explanations for a perceived race difference in athleticism: The relation to prejudice toward and stereotyping of blacks." *Athletic Insight: Online Journal of Sport Psychology* 9(3).

Sheriff, R. E. (2001). *Dreaming equality: Color, race, and racism in urban Brazil*. New Brunswick, NJ, Rutgers University Press.

Shih, M., N. Ambady, J. A. Richeson, K. Fujita, and H. M. Gray (2002). "Stereotype performance boosts: The impact of self-relevance and the manner of stereotype activation." *Journal of Personality and Social Psychology* 83: 638–64.

Shih, M., T. L. Pittinsky, and N. Ambady (1999). "Stereotype susceptibility: Identity salience and shifts in quantitative performance." *Psychological Science* 10: 80–3.

Shoemaker, D. (2015). *Responsibility from the margins*. Oxford, Oxford University Press.

Shulman, J. L. and J. Glasgow (2010). "Is race-thinking biological or social, and does it matter for racism? An exploratory study." *Journal of Social Philosophy* 41(3): 244–59.

Slater, M. H. (2015). "Natural kindness." *British Journal for the Philosophy of Science* 66(2): 375–411.

Smart, J. J. C. (1959). "Sensations and brain processes." *Philosophical Review* 68(April): 141–56.
Smedley, A. and B. D. Smedley (2012). *Race in North America: Origin and evolution of a worldview*. Boulder, CO, Westview Press.
Snowden, F. M. (1983). *Before color prejudice: The ancient view of Blacks*. Cambridge, MA, Harvard University Press.
Sober, E. (1980). "Evolution, population thinking, and essentialism." *Journal of Philosophy* 47: 350–83.
Solomon, G. E. A. (2002). "Birth, kind, and naive biology." *Developmental Science* 5(2): 213–18.
Sosa, E. (1993). "Putnam's pragmatic realism." *Journal of Philosophy* 60(12): 605–26.
Spanos, N. and C. Burgess (1994). Hypnosis and multiple personality disorder: A sociocognitive perspective. In *Dissociation: Clinical and theoretical perspectives*. S. J. Lynn and J. W. Rhue. New York, Guilford Press: 136–58.
Spencer, Q. (2014). "A radical solution to the race problem." *Philosophy of Science* 81(5): 1025–38.
Sperber, D. (1994). The modularity of thought and the epidemiology of representations. In *Mapping the mind: Domain specificity in cognition and culture*. L. A. Hirschfeld and S. A. Gelman. New York, Cambridge University Press: 39–67.
Sperber, D. (1996). *Explaining culture*. Cambridge, MA, Blackwell Publishers.
Springer, K. and F. C. Keil (1989). "On the development of biologically specific beliefs: The case of inheritance." *Child Development* 60: 637–48.
Springer, K. and F. C. Keil (1991). "Early differentiation of causal mechanisms appropriate to biological and nonbiological kinds." *Child Development* 62: 767–81.
Sripada, C. and S. Stich (2006). A framework for the psychology of norms. In *The innate mind: Culture and cognition*. P. Carruthers, S. Laurence, and S. Stich. New York, Oxford University Press: 280–301.
Stalnaker, R. (1989). "On what's in the head." *Philosophical Perspectives* 3: 287–316.
Stanford, P. K. and P. Kitcher (2000). "Refining the causal theory of reference." *Philosophical Studies* 97(1): 99–129.
Stanovich, K. E. (2004). *The robot's rebellion: Finding meaning in the age of Darwin*. Chicago, IL; London, University of Chicago Press.
Staples, B. (1986). "Just walk on by." *Ms.* 15(3): 54.
Steele, C. and J. Aronson (1995). "Stereotype threat and the intellectual test performance of African-Americans." *Journal of Personality and Social Psychology* 69: 797–811.
Stein, E. (1990). Conclusion: The essentials of constructionism and the construction of essentialism. In *Forms of desire: Sexual orientation and the social constructionist controversy*. E. Stein. New York, Routledge: 325–53.
Stein, E. (1999). *The mismeasure of desire*. New York, Oxford University Press.
Steinpreis, R. E., K. A. Anders, and D. Ritzke (1999). "The impact of gender on the review of the curricula vitae of job applicants and tenure candidates: A national empirical study." *Sex Roles* 41(7/8): 509–28.
Sterelny, K. (2012). *The evolved apprentice: How evolution made humans unique*. Cambridge, MA, MIT Press.

Sterelny, K. and P. Griffiths (1999). *Sex and death: An introduction to philosophy of biology*. Chicago, IL, University of Chicago.

Stich, S. P. (1983). *From folk psychology to cognitive science: The case against belief*. Cambridge, MA, MIT Press.

Stich, S. (1996). Deconstructing the mind. In *Deconstructing the mind*. New York, Oxford University Press: 3–90.

Stoljar, N. (1995). "Essence, identity and the concept of woman." *Philosophical Topics* 23: 261–93.

Stone, J. (2002). "Battling doubt by avoiding practice: The effect of stereotype threat on self-handicapping in white athletes." *Personality and Social Psychology Bulletin* 28: 1667–78.

Stone, J., C. I. Lynch, M. Sjomeling, and J. M. Darley (1999). "Stereotype threat effects on black and white athletic performance." *Journal of Personality and Social Psychology* 77: 1213–27.

Strawson, P. (1962). "Freedom and resentment." *Proceedings of the British Academy* 48: 1–25.

Strevens, M. (2000). "The essentialist aspect of naive theories." *Cognition* 74: 149–75.

Strevens, M. (2001a). "Reply to Ahn et al." *Cognition* 82: 71–6.

Strevens, M. (2001b). "Further comments on Ahn et al." Retrieved January 5, 2012, from <http://www.strevens.org/research/cogsci/ahnetal.pdf>.

Sundstrom, R. (2002). "Racial nominalism." *Journal of Social Philosophy* 33: 193–210.

Sundstrom, R. (2003). "Race and place: Social space in the production of human kinds." *Philosophy and Geography* 6(1): 83–95.

Szasz, T. S. (1974). *The myth of mental illness: Foundations of a theory of personal conduct*. New York, Harper and Row.

Tajfel, H. (1970). "Experiments in intergroup discrimination." *Scientific American* 223 (November): 96–102.

Taylor, C. (1971). "Interpretation and the sciences of man." *Review of Metaphysics* 25: 3–51.

Taylor, C. (1976). Responsibility for self. In *The identities of persons*. Amélie Rorty. Berkeley, CA, University of California Press: 281–99.

Taylor, P. (2000). "Appiah's uncompleted argument: Du Bois and the reality of race." *Social Theory and Practice* 26(1): 103–28.

Taylor, P. C. (2013). *Race: A philosophical introduction*. Cambridge; Malden, MA, Polity Press.

Teachman, B. A. and K. D. Brownell (2001). "Implicit anti-fat bias among health professionals: Is anyone immune?" *International Journal of Obesity* 2: 1525–31.

Telles, E. (2002). "Racial ambiguity among the Brazilian population." *Ethnic and Racial Studies* 25(3): 415–41.

Templeton, A. R. (2013). "Biological races in humans." *Studies in History and Philosophy of Science Part C: Studies in History and Philosophy of Biological and Biomedical Sciences* 44: 262–71.

Thomasson, A. (2003). "Realism and human kinds." *Philosophy and Phenomenological Research* 67(3): 580–609.

Thomasson, A. L. (2007). *Ordinary objects*. New York, Oxford University Press.

Thornhill, R. and C. T. Palmer (2000). *A natural history of rape: Biological bases of sexual coercion*. Cambridge, MA, MIT Press.

Thornhill, R. and N. W. Thornhill (1992). "The evolutionary psychology of men's coercive sexuality." *Behavioral and Brain Sciences* 15(2): 363–75.

Thornhill, R. and N. W. Thornhill (1992). "The study of men's coercive sexuality: What course should it take?" *Behavioral and Brain Sciences* 15(2): 404–21.

Tomasello, M. (2008). *Origins of human communication*. Cambridge, MA, MIT Press.

Tooby, J. and L. Cosmides (1992). The psychological foundations of culture. In *The adapted mind*. J. Barkow, L. Cosmides and J. Tooby. New York, Oxford University Press: 19–136.

Toynbee, A. J. (1957). *A study of history, Vol 2: Abridgment of vols 7–10*. Ed. D. C. Somervell. Oxford, Oxford University Press.

Trivers, R. (1976). Foreword. In *The selfish gene*. R. Dawkins. New York, Oxford University Press: v–vii.

Tsou, J. Y. (2007). "Hacking on the looping effects of psychiatric classifications: What is an interactive and indifferent kind?" *International Studies in the Philosophy of Science* 21(3): 329–44.

Tye, M. (1998). "Externalism and Memory." *Aristotelian Society* Supp.

Tyson, K. (2011). *Integration interrupted: Tracking, black students, and acting white after brown*. New York, Oxford University Press.

Tyson, K., W. Darity, Jr., and D. R. Castellino (2005). "It's not 'a black thing': Understanding the burden of acting white and other dilemmas of high achievement." *American Sociological Review* 70(4): 582–605.

Valian, V. (1998). *Why so slow? The advancement of women*. Cambridge, MA, MIT Press.

Van der Hart, O. (1993). "Multiple personality disorder in Europe: Impressions." *Dissociation* 6(2/3): 102–18.

van Fraassen, B. C. (1980). *The scientific image*. New York, Clarendon Press; Oxford University Press.

Vanderschraaf, P. and G. Sillari. (2009). "Common knowledge." In *Stanford encyclopedia of philosophy*. Retrieved March 7, 2014, from <http://plato.stanford.edu/archives/spr2009/entries/common-knowledge/>.

Velleman, J. D. (2000). *The possibility of practical reason*. New York, Oxford University Press.

Verplaetse, J., S. Vanneste, and J. Braeckman (2007). "You can judge a book by its cover: The sequel. A kernel of truth in predictive cheating detection." *Evolution and Human Behavior* 4: 260–71.

Vohs, K. and J. W. Schooler (2008). "The value of believing in free will: Encouraging a belief in determinism increases cheating." *Psychological Science*. 19(1): 49–54.

von Hippel, W. and R. Trivers (2011). "The evolution and psychology of self deception." *Behavioral and Brain Sciences* 34: 1–56.

Vrij, A. and S. Mann (2005). Police use of nonverbal behavior as indicators of deception. In *Applications of nonverbal communication*. R. E. Riggio and R. S. Feldman. Mahwah, NJ, Erlbaum: 63–94.

Walzer, M. (1983). *Spheres of justice: A defense of pluralism and equality.* New York, Basic Books.

Washington, N. and D. Kelly (forthcoming). Who's responsible for this? Moral responsibility, externalism, and knowledge about implicit bias. In *Implicit bias and philosophy.* M. Brownstein and J. Saul. Oxford, Oxford University Press.

Weiner, B., R. P. Perry, and J. Magnusson (1988). "An attributional analysis of reactions to stigma." *Journal of Personality and Social Psychology* 55: 738–48.

Williams, M. J. and J. L. Eberhardt (2008). "Biological conceptions of race and the motivation to cross racial boundaries." *Journal of Personality and Social Psychology* 94(6): 1033–47.

Wilson, D. S. (2005). Evolutionary social constructivism. In *The literary animal: Evolution and the nature of narrative.* J. Gottschall and D. S. Wilson. Evanston, IL, Northwestern University Press: 20–37.

Wilson, D. S., E. Dietrich, and A. B. Clark (2003). "On the inappropriate use of the naturalistic fallacy in evolutionary psychology." *Biology and Philosophy* 18(5): 669–81.

Wilson, M. and M. Daly (1992). The man who mistook his wife for a chattel. In *The adapted mind: Evolutionary psychology and the generation of culture.* J. H. Barkow, L. Cosmides, and J. Tooby. New York, Oxford University Press: 289–322.

Wilson, R. A. (1999). Realism, essence, and kind: Resuscitating species essentialism? In *Species: New interdisciplinary essays.* R. A. Wilson. Cambridge, MA, MIT Press.

Wilson, R. A., M. J. Barker, and I. Brigandt (2007). "When traditional essentialism fails." *Philosophical Topics* 35(1–2): 189–215.

Wilson, T. D. (2002). *Strangers to ourselves: Discovering the adaptive unconscious.* Cambridge, MA, Belknap Press of Harvard University Press.

Witt, C. (1995). "Anti-essentialism in feminist theory." *Philosophical Topics* 23: 321–44.

Witt, C. (2011). *The metaphysics of gender.* New York, Oxford University Press.

Wittgenstein, L. (1958). *Philosophical investigations*, trans. G. E. M. Anscombe. New York, Blackwell.

Wollstonecraft, M. (1995). A vindication of the rights of woman. In *A vindication of the rights of man with a vindication of the rights of woman and hints.* S. Tomaselli. New York, Cambridge University Press: 65–294.

Woolfolk, R. L., J. M. Doris, and J. M. Darley. (2006). "Identification, situational constraint, and social cognition: Studies in the attribution of moral responsibility." *Cognition* 100(2): 283–301.

Wright, C. (1992). *Truth and objectivity.* Cambridge, MA, Harvard University Press.

Zack, N. (1993). *Race and mixed race.* Philadelphia, PN, Temple University Press.

Zack, N. (1999). "Philosophy and racial paradigms." *Journal of Value Inquiry* 33: 299–317.

Zack, N. (2002). *Philosophy of science and race.* New York, Routledge.

Zinn, H. (2005). *A people's history of the United States.* New York, Harper Perennial Modern Classics.

Index of Names

Ahn, W. 29, 43
Akins, Kathleen 98
Alcoff, Linda 68n1, 214
Amodio, D. 78
Anderson, Elizabeth 88
Andreasen, Robin 23, 39
Anscombe, G.E.M. 54
Appiah, Kwame Anthony 23, 24, 28, 68n1, 70–1, 111, 115, 175
Aristotle 162
Aronson, Joel 79, 80
Aspinwall, Lisa 105
Ástá Sveinsdóttir 88n16, 211–13
Astuti, Rita 25, 40–2
Atran, Scott 25, 30, 75
Auster-Gussman, Lisa 106
Averill, James 49, 70, 96, 111, 118

Bach, Theodore 92n20
Banton, Michael 15, 21–2, 23, 28
Bargh, John 80
Baron, Andrew Scott 76–7
Barrett, H. Clark 30, 75
Barzun, J. 23n11
Baumeister, Roy 107
Bayly, Susan 32, 34
Bellows, Nancy 125
Berger, Peter 146
Bermúdez, José 180
Birnbaum, D. 42
Bloor, David 145–6
Blum, Lawrence 23
Boehm, Christopher 178, 179
Bourdieu, Pierre 82n14
Boyd, Richard 26n14, 69, 90–3, 139, 144, 146, 147, 150, 156, 186, 198, 201, 203
Boyd, Robert 44, 45, 77, 178, 179
Boyer, Pascal 174n12
Brandon, R.N. 69
Brosig, J. 119
Brownell, K.D. 78, 79
Bruner, J.L. 142
Buffon, G.L.L. 21
Buller, David 108n13
Burnette, Jeni 106
Buss, David 97, 127
Butler, Judith 5, 10, 48, 70, 111, 115, 149

Carey, Susan 25, 76–7
Cesario, Joseph 80n10
Chartrand, Tanya 80n10
Chengtian, He 33
Chudek, M. 75, 178, 179
Churchland, Paul 139, 187
Claire, T. 79
Cohen, Harold 39
Collins, Harry 143
Condit, C.M. 26, 27
Cosmides, Leda 77, 97, 117
Craver, Carl 93n21
Croizet, J. 79

Daly, Martin 97, 108n13
Darley, John 105
Davidson, Arnold 6, 21, 49, 55, 88
Davidson, Joyce 214n1
Dawkins, Richard 119
DePaulo, B. 119
Devine, P. 78
Devitt, Michael 6n6, 23, 148, 155n17, 169n8, 188, 196, 198, 200
DeWall, C. Nathan 107
Dikötter, Frank 32, 33
Doris, John 105, 122n8, 123, 129–30
Doyen, S. 80n10
Drabek, Matt 68–9n1
Dray, W.H. 69
Dummett, Michael 147, 155n18
Dunham, Yarrow 76–7, 78
Dupré, John 98, 100–1, 108
Dweck, Carol 73, 175

Eberhardt, Jennifer 27n16
Egan, Andy 150
Ekman, Paul 81, 97, 176
Ereshefsky, Marc 92n20
Eshleman, A. 132n15
Evans, Gareth 187, 194–5

Fair, D. 151
Faucher, Luc 17, 25, 30
Fehr, Ernst 81
Festinger, Leon 120n5
Feyerabend, Paul 10
Field, Hartry 194n8, 215n2
Fine, Cordelia 108–9
Fischer, John Martin 131–2

INDEX OF NAMES

Fiske, A.P. 174n12
Fodor, Jerry 43–4n34, 47, 90, 139
Fordham, Signithia 74, 184–5
Foucault, Michel 2, 5, 8, 21, 49, 55, 65, 68n1, 141, 191, 210
Frank, R. 119
Frantz, C.M. 79
Fredrickson, George 15, 16, 20, 22, 23, 28, 33, 35, 38–9, 47n35
Fricker, Miranda 84
Friesen, Wallace 81, 176
Fryer, Roland 74, 184–5

Gannett, Lisa 23, 24, 208, 213
Geertz, Clifford 52
Gelman, Susan 17, 25–7, 28, 29, 97, 200
Gendler, Tamar 7, 123
Gibson, C.E. 79–80n9
Gilbert, Margaret 61, 64n17, 211
Gil-White, Francisco 17, 25, 30, 31n24, 34, 37, 39, 179, 180
Giner-Sorolla, R. 107n12
Ginet, Carl 132n15
Glasgow, Joshua 23, 27, 27n16, 194n7, 208
Gobineau, A. 21
Godfrey-Smith, Peter 30n21
Goffman, Erving 49, 111
Gonzales, P.M. 79
Gooding-Williams, Robert 26
Goodman, Nelson 149
Greene, Joshua 77, 179
Greenwald, Anthony 78, 79n8
Griffith, Aaron 155n19
Griffiths, Paul 17n4, 58n5, 68n1, 81, 90, 92–3, 111, 117, 118, 121–2, 169n8, 200, 211, 216
Guillaumin, Colette 15, 22, 23

Hacking, Ian 5–6, 10, 51, 53–8, 65–6, 68n1, 69–71, 72, 111, 114, 156–7, 163, 164, 169–73, 175, 177, 179, 211
Haidt, Jonathan 3
Hall, Stuart 2, 48, 181
Hannaford, Ivan 15–16
Hardimon, Michael 23, 168n6
Harman, Gilbert 61, 62n10
Harris, Angel L. 74, 184–5
Harris, Marvin 35–40
Harwood, Jonathan 15, 21–2, 23, 28
Haslam, N. 27n16, 44, 45
Haslanger, Sally 23, 68n1, 109n15, 128, 186, 209–10, 211
Heidegger, Martin 82n14
Henrich, Joseph 32, 75, 77, 178, 179
Hieronymi, Pamela 110n16
Hirschfeld, Lawrence 17, 25, 26n14, 28–9, 30–1, 38, 97

Hoff, Karla 81
Hood, Gloria 105
Hoyt, Crystal 106, 109

Isaac, Benjamin 32, 33
Izard, Carol 81

Jackson, Frank 19, 51, 187
Jayaratne, T.E. 27
Jenkins, C.S. 148
Johnson, S.C. 28
Johnston, A.I. 33
Jones, Doug 17, 25, 31n23, 37–40

Kamtekar, Rachana 33
Kanovsky, Martin 25, 29n19, 34–5
Kaplan, Jonathan Michael 87
Keil, Frank 25, 26, 29, 30, 75, 200
Kelly, Dan 9, 75, 77, 79n8, 108, 147, 178
Khalidi, Muhammad Ali 6n6, 93n21, 200
Kinzler, Katherine 31, 77
Kitcher, Philip 23, 39, 139, 146, 178–9, 188n2, 196–7, 198–201
Klein, Peter 62n9
Kornblith, Hilary 93n21
Kottack, Conrad 36
Krebs, John 119
Kripke, Saul 26n14, 53n3, 90, 185, 196
Kshetramande, Mayuresh 81
Kuhn, Thomas 10, 142, 143–4, 149, 153, 184
Kunda, Ziva 129n12
Kuorikoski, J. 68n1
Kurth, Charlie 178
Kurzban, Robert 77, 179

Laland, Kevin 82n14
Laqueur, Thomas 141, 142
Latour, Bruno 143
Lazarus, R. 52
Lenski, Richard 167
Leslie, Alan 75, 178
Leslie, Sarah-Jane 28n17, 77
Lewis, David 59, 61, 62, 64n17, 150, 151, 176, 187, 211
Longino, Helen 146
Luckmann, Thomas 146
Lutz, Catherine 52
Lycan, William 187

Machery, Edouard 17, 18n6, 25, 30, 43–4n34, 45, 108, 188, 192
MacKinnon, Catherine 2, 8, 48, 111, 128–9, 130
Magnussen, Jamie 105
Mahalingam, R. 42
Maibom, Heidi 180
Malle, Bertrand 106

INDEX OF NAMES 243

Mann, S. 120
Markman, Ellen 25, 27
Martin, C.L. 27
Marx, Karl 141–2
Masicampo, E.J. 107
Mayr, Ernst 18n6, 23, 91–2, 200
McElreath, Richard 78, 178, 179
McKenna, Michael 132n15
McKenna, Wendy 7
Medin, Douglas 25, 75
Mercier, Hugo 3
Merton, Robert 80
Miller, Alexander 148, 154–5
Millikan, Ruth 92n20, 169n8
Mills, Charles 23, 142, 158
Monahan, John 105
Monterosso, John 105
Moon, A. 79–80n9
Moya, C. 44, 45
Murphy, Dominic 68n1, 71n2, 175n13
Murray, Dylan 105

Nahmias, Eddy 105, 122n8, 126, 130
Nichols, Shaun 147, 188, 208
Nicholson, Linda 214n1
Nisbett, Richard 124–6
Nogueira, Oracy 35–6, 38
Nolan, J.M. 125
Nosek, Brian 78, 79

Ogbu, John Uzo 74, 184–5
Omi, Michael 48, 162–3
Oswald, F.L. 79n8
Outlaw, Lucius 23

Palmer, Craig 97–8
Parker, S. 27
Pashler, H. 80n10
Payne, Keith 79
Perry, Raymond 105
Petersen, M.B. 174n12
Pickering, Andrew 141, 142
Pietraszewski, D. 77, 179
Pinch, Trevor 143
Pinillos, Ángel 188n2
Pinker, Steven 98, 99, 129
Piper, Adrian 26
Plato 94, 96n4, 146, 162, 194n6
Pöyhönen, S. 68n1
Putnam, Hilary 26n14, 47, 90, 147, 149, 185, 196, 197

Quine, W.V.O. 20

Ravizza, Michael 131–2
Rhodes, Marjorie 77
Richerson, Peter 77, 178, 179

Riecken, Henry 120n5
Robbins, Philip 122n7
Robinson, E. 35
Rodriguez, J. 42
Roeder, S. 79–80n9
Root, Michael 23
Rorty, Richard 149
Rosen, Gideon 138, 149–150, 152–3
Royzman, Edward 105

Salmon, W. 151
Sartre, Jean-Paul 70, 96, 103–4, 111, 115
Sayre-McCord, Geoffrey 158
Schachter, Stanley 120n5
Schaffer, Jonathan 155n19
Scheff, Thomas 49, 111, 171
Schneider, David 27
Schooler, Jonathan 107
Schroeter, Laura 197, 200, 202
Schwartz, Barry 105
Searle, John 63n14, 88n16, 191–4, 211
Sesardic, Neven 23
Shackelford, T.K. 127
Shafer-Landau, Russ 158
Shariff, Azim 105, 109
Sheldon, J.P. 27
Sheriff, R.E. 35
Shih, M. 79
Shulman, Julie 27, 27n16
Sillari, G. 60n7
Slater, Matthew 93n21
Smart, J.J.C. 156
Smedley, Audrey 16, 23n11, 32n26
Smedley, Brian 16, 23n11, 32n26
Smith, Mick 214n1
Snowden, F.M. 33
Sober, Elliot 18n6, 23
Solomon, G.E.A. 28, 29
Sosa, Ernest 148
Spencer, Quayshawn 23
Sperber, Dan 3, 17, 39, 174
Springer, Ken 25
Sripada, Chandra 75, 178
Stanford, P. Kyle 188n2, 196–201
Stanovich, Keith 80
Staples, Brent 172
Steele, Claude 79, 80
Stein, Edward 94n1, 110n17
Sterelny, Kim 82n14, 83, 188, 196, 198, 200
Stich, Stephen 52–3, 75, 139, 178, 187
Stoljar, Natalie 214n1
Stone, J. 79
Strawson, Peter 100, 103–4, 131–2
Strevens, Michael 43
Sundstrom, Ronald 88
Szasz, Thomas 49

Tajfel, Henri 76
Taylor, Charles 130, 162–6, 170, 171, 173, 177
Taylor, Paul C. 23, 47n35, 48
Teachman, B.A. 78, 79
Telles, E. 39
Thomasson, Amie 192–3, 196
Thornhill, Nancy 97–101, 108
Thornhill, Randy 97–101, 108
Tooby, John 77, 117
Torelli, Paul 74, 184–5
Toynbee, Arnold 143
Trivers, Robert 121, 122
Tworek, Christina 77
Tyson, K. 74, 184–5

Valian, Virginia 79n8
Vanderschraaf, Peter 60n7
van Fraassen, Bas 139n4, 146n13
Vellemen, David 122
Verplaetse, J. 119

Vohs, Kathleen 107
von Hippel, W. 122
Vrij, A. 120

Walzer, Michael 171–2
Weinberg, Jonathan 17n4
Weiner, Bernard 105
Wellman, Henry 25, 28
Williams, M.J. 27n16
Wilson, Margo 97, 108n13
Wilson, Robert 90–1, 93, 169n8
Wilson, Timothy 124–6
Winant, Howard 48, 162–3
Windham, Mary 98
Wittgenstein, Ludwig 214n1
Wollstonecraft, Mary 128
Woolfolk, Robert 105
Woolgar, Stephen 143

Zack, Naomi 23, 40n32, 108, 211
Zinn, Howard 94–5, 96, 99, 108

Index of Subjects

accent preference 31, 77
acquired automaticity 80–1, 82
  as adaptation 117n3
  see also automatic processes
"acting white" 74–5, 184–5
action
  individuated by description 54–6
  intentional
    as mechanism of category construction 68, 69–75
    as rational 69–71
    as stabilizing of category 175
action analysis of category construction 50, 69, 71–2
  as necessary for category 54–5
  shortcomings of 55–8
  as sufficient for category 56
agency 10, 112–13, 129–33, 210
anti-psychiatric views 49
anti-realism
  global 137, 156n21
  local 138
  see also realism
austere descriptions *see* descriptivism
automatic processes
  characterized 75
  dissociated from intentional processes 79
  implicit biases 78
  as mechanism of category construction 57, 68, 69, 75–81
  as mechanism of category stability 176
  see also acquired automaticity

bad faith 103–4
  collective bad faith model of performance 120–1
basic realism 140
  see also realism
Bayesian learning algorithms 18n7
biobehavioral kinds 50, 163, 189, 207
Brazil, racial thinking in 35–40
  see also evolutionary cognitive account of essentialism, hard cases for; racial essentialism
broadcast *see* public broadcast
broad essentialism
  characterized 25–6
  about race 27–8

canalized, culturally 17–18, 25, 30, 42, 44, 45–6
  beliefs about essence, vs. perceptually acquired 30–2
  as changeable 18, 81n12
  of whatever mechanisms explain inferences about race 44
caste 34, 45, 81, 171–2
categories
  biological naturalist views of 2–3
  characterized 6, 6n6
  constructionist views of 2–3
  realism versus skepticism regarding 2–3, 182–4
  see also kinds
category construction 2–3
  as causally significant 50, 55, 69, 211
  challenges for 50
  characterized 7–8, 49, 147
  as competing with nonconstructionist explanation 160
  general versus particular mechanisms of 68
  versus representations 1, 6, 6n6, 47
    asymmetric plausibility 9
    differing threats to realism 141
causal essentialism 26n14
The causal explanatoriness of k-hood 29
  see also inheritance thinking
causal historical reference 144, 185, 202–4
  adoption by social constructionists 185–6
  allows mistaken description 186
  as austere account of meaning 53n3
  characterized 187
causally homeostatic property clusters *see* homeostatic property cluster kinds
The causal necessity of parent kind 29
  see also inheritance thinking
The causal sufficiency of parent kind 29
  see also inheritance thinking
child abuse 169–70, 171, 172
coalitions 77
cognitive dispositions
  conceptual change possible 174
  as stabilizing 173–5
  see also racial essentialism
collateral reference 185, 203
collective acceptance 191, 192
collective action problem 215–16
collective bad faith model *see* bad faith

collective belief 62
common knowledge 58, 70, 82, 89, 93, 175, 207–8
 as background cause of behavior
  via automatic processes 80
  via intentional action 73–5, 116, 124, 127
 characterized 59–60
 as idealization 62–3
 iterated statement of 61–2
 kernel of need not be true 59n6
 and network effects 61
 produced by public broadcast 64–6
 versus shared belief 60
 stronger condition than required 80n11
computational unconscious 122
Conceptual Break Hypothesis
 characterized 16–24
 externalist interpretation of 47, 204
 as a failure of conceptual identity 16
  see also HERE hypothesis; racial essentialism
 versus nonconceptual break 19, 46–7
conferralist account of social construction 211–13
Confucian and Mencian emphasis on the unity of humankind 33–4
conspicuous indicator
 of category membership 66
 of social role membership 180
 volitional versus nonvolitional 66
The Construction of Human Kinds 8
co-reference 144, 184, 203
covert category construction 207
 characterized 49
 contrasts with biobehavioral explanation 49–53
 not conventional or institutional kinds 63, 211
 problem with reference 182–97
 as target of performative constructionists 118
critical constructionists 142
cultural networks 83–4
cultural transmission
 as possible source of racial essentialism 30–1, 42
 as requiring an "origination" story 30n22

debunking 46, 96–7, 172–3
 as constructionist threat 140–1, 143, 144–7
 global vs. local 146–7
deception 119–20
deflationary critique of the evolutionary-cognitive account of essentialism 43–4
descriptivism
 versus causal historical approaches 185–6
 characterized 19, 187

descriptions in
 austere vs. opulent construals 20, 51–3, 56
 core vs. peripheral elements 19–20
 holistic vs. molecular construals 19–20
 difficult to distinguish meaning constitutive elements 20, 51
desires
 construction of 7
 instrumental vs. primitive 125, 129
developmentally invariant *see* canalized, culturally
discursive formations 5–6, 210
display rules 81
dissociation 122
domain-specific mechanism 17–18, 42, 45
 vs. Bayesian learning algorithms 18n7
 proper domain 39
Dual Constructionism 7, 9
dynamic nominalism 5–6, 204, 210

eliminativism 2–3, 187
emotions
 anger 52
 basic 81
 construction of 49, 96, 121–2, 125
 sadness 103, 115
environmental construction 57–8, 82–9, 213
 characterized 82–3
 as constructing categories 88–9
 as mechanism of category construction 68, 69
 as stabilizing categories 176
epistemic constraint 83, 140, 145, 170
 characterized 165
epistemic weighting 84
essence of a natural kind 90–2
 see also homeostatic property cluster kinds; kinds
essentialist thinking 174, 199–200, 201, 208
 about biological entities 9, 24–30
 biologically false 9, 23, 183, 200
 about human kinds (broad and lineage) 77
 see also broad essentialism; evolutionary cognitive account of essentialism; lineage essentialism; nonessentialism; racial essentialism
evolutionary cognitive account of essentialism
 characterized 16–19, 24–5
 critiques of 43–6
 cross-cultural evidence for 32–42
 as debunking 46
 evolutionary cognitive theorists as racial skeptics 18n5
 evolutionary mismatch to human groups 39–40
 experimental evidence for 24–32
 hard cases for, in case of race 32, 35
 no race module 17

transferred from biological to racial domains 32, 39–40, 39n31, 41–2
versus cultural transmission of theory 30–2
versus perceptually acquired 30–2
evolutionary cognitive adaptations 97
explanation-driven metaphysics 208
explanatory particularism 68
explicit judgments 76
externalism about reference *see* causal historical reference

failed reference 182–4
failures of self-knowledge 118–33, 175, 208
failure to detect and identify 121–3, 129–30
failure to locate model 10, 112, 113, 126–32
family resemblance concepts 214n1
folk biology 25–30
folk theories, characterized 25n13
Four-Way Stop 60–2
fugue 170

gender and sex 141
construction of 2, 48–9, 70
as female passivity 128–9, 130–1
as a genuine kind 6
as performed 48, 111, 112, 115
as referent of gender terms 8
as sex difference 193–4
as sexual infidelity 94, 126–8
as a thin natural kind 189–90
group beliefs 7

Hacking Instability 169–71, 173
characterized 171
versus *Taylor Instability* 170
HERE hypothesis 17, 18, 20–4, 28, 36, 42, 45–7
alternatives to 46
characterized 21
see also racial essentialism
homeostatic property cluster (HPC) kinds 69, 89–93, 166, 198, 207
alternatives 167–9
characterized 90–2
coinstantiation of properties in need of explanation 201
mechanism requirement too demanding 93n21
neutral with respect to character of homeostatic mechanism 92
possibility of historic individuating properties 92n20, 169n8
understood a posteriori 213, 214
see also kinds
homosexuality 21, 49, 55, 70, 141
construction of 11–12
as a genuine kind 6
social role produced by public broadcast 65

how possibly account 5, 8, 63, 68–9, 116, 207
human natural history 178–80
human naturalism 152–3, 211
"human nature wars" 160
hybrid constructionism 9, 175
hybrid theories of reference 187, 196
hypodescent 28n18, 39, 42

identification 57
ideology 142
implicit attitudes 7
characterized 78
Implicit Association Test (IAT) 78–9
implicit measures 76–7
see also automatic processes
imposition of function 191
incommensurability arguments 143–4
inheritance thinking
as an element in historical representations of race 28
characterized 29
in children 28–9
innate *see* canalized, culturally
instability (of categories) *see* stability, of categories
institutional kind 189, 191–4, 202, 203
not a covert construction 192–4
see also kinds
institutions
fixing ascription conditions 85
see also hypodescent
fixing norms 85–6
material transformation 86–8
intentional action *see* action
intersectional categories 215

justice-driven metaphysics 210

kinds
characterized 6
conceptions of 167–9, 204
essential features
combined mechanisms 204
intrinsic versus relational characterizations of 196–7, 199–200, 202
human kinds "on the move" 163, 164, 166, 172
interactive versus indifferent 157
interest relative 6, 67, 89–90
natural 2, 11, 26n14, 50, 72, 85, 89–90, 182
thin 189–91, 202
social 182
stability of 162–81
existence versus features 168
various factors in 173
see also homeostatic property cluster kinds; institutional kind

INDEX OF SUBJECTS

labeling effects 169–71, 173
   characterized 171
learning scaffolds 83, 84–5
lineage essentialism 28–30, 32–42
   cross-cultural evidence for, in representations of human groups 33–42
   pairing of broad essentialism and inheritance thinking 29
   representation of as a cognitive adaptation 30, 39–40
   see also inheritance thinking; racial essentialism
Literal 137, 138, 140–1
looping effect of human kinds 70, 156–7, 163, 169
   feedback loops as stabilizers 177

making up people 53–8, 70, 111, 191
   see also action analysis of category construction
malingering 113–14
   model of performance 119–21
Martian anthropologist 154, 159–60, 215
Martian robot cats *see* robot cats, Martian
meaning change
   characterized 19–20
   constituting conceptual break 16
   externalist approach to 47
   see also descriptivism
mere distinction *see* minimal groups
metaphysically moderate construction 11, 137, 154–5
Michael the Philanderer 127, 130
mindset, growth versus fixed 73
minimal groups 76–8, 179
minimal hypothesis, instead of folk essentialism 43
mismatch problem 183–4
   returns for grounding 197
missing referent, problem of the 186, 189, 202
mixed-race 37–40, 40n32, 109
moral hazard 100, 210
   characterized 95
   production of by some representations 106–7
   roles in constructionist theorizing 95–6
motivated cognition 129n12
multiple personality 55, 175
   as iatrogenic 65
   as an intentional response to social role 70, 114
      belief-desire model 123–4
   precursors of 56
   social role produced by public broadcast 65–6
mundane dependence 148
   causal demarcation 149–52
   as nonfundamental 155–6
Myth of the Metals 94, 96n4

Nash equilibrium 63, 175–6
natural, representation as 72, 73, 94–107
   cultural variability in 104n11
   production of moral hazard 106–7
   reduction of reactive attitudes 105–6
Naturalism 149
   characterized 4, 102
   as conception of human mind 152
   committed independence of world 159–60
naturalistic fallacy 97–101
   characterized 98–9
naturalists 149
   biological racial 3, 21
   as constructionist critics 97, 129
   understood as realists about categories 2–3
Necessary-Description Category Construction 8, 16, 53–6
neo-Kantian Construction 147, 149–54
Neurosemantic Eraser Ray 89
nominalist 211
nomological dangler 156
nonessentialism 213–14
nonintentional processes *see* automatic processes
Nonstrategic reasons 71–3
norms 104n10
   adaptations to conform 178
   as cognitive aids for living in large groups 179

obesity 106, 109
objective attitude 103, 131–2
Objectivity 137, 138, 140–1, 147–54
   constructionist threat to 141
   as mind independence 147–8, 153
   see also mundane dependence
"one-drop of blood" rule *see* hypodescent
opulent baptism 197, 200
opulent descriptions *see* descriptivism

partial construction 185, 203, 213
partial reference 194n8, 215n2
passing 26
performative constructionist explanations 10, 70, 111–29
   characterized 113–17
perversion 49, 88, 183
political animal 3, 162
populations, human 17, 19, 30, 39–40
proper domain 39–40, 39n31
propositional attitudes *see* representational attitudes
proximally intentional 116–17
public broadcast 64–6, 84, 93
puzzle of intention and ignorance 10, 112, 118, 124

qua problem 195–202
    characterized 195–6
    description as solution to 196, 199
quietism 154

race
    construction of 2, 48, 70
    as a genuine kind 6
    naturalism about 3
    skepticism about 3
    as a thin natural kind 189, 191
    as unstable 163
racial essentialism 18–19, 199
    as allowing for passing 26
    assumptions about outrun perceptual
        capacities 30–2, 31n23
    biologically false 23, 42, 200
    characterized 16–17
    conception of as consequence of cognitive
        predisposition 24–32
    conception of as historically particular
        16–17, 20–4
    cross-cultural evidence of belief in 33–42
        Brazil 35–40, 42
        China 33–4
        premodern-Europe and the
            Mediterranean 33
        India 35
        Mongolia 34, 42
        Ukraine 34–45
        Vezo of Madagascar 40–2
    as lineage essentialism about race 29
    variations in belief in 32–3, 35, 35–42
racial identity 70, 74, 115
racialism see racial essentialism
racial representations
    constructionist explanation of 15, 45–6
    content of
        determined by many factors 42, 45–6
        as determined by referent 47
        "in blood" 22, 28, 29–30, 38, 39
        "inborn" or "endogenous" 22–3, 28, 30, 33
        parallel to biological kind representations
            22, 24–30, 32
        as passed on in biological reproduction
            28–9
        shifting from biological to cultural
            explanations 27
        skin as criterial 52
    evolutionary cognitive explanation of 17–18,
        45–6
    invented 9–10, 15
    used for group generalizations 27–8
    see also racial essentialism
racial skepticism 183–4
    as consistent with an evolutionary cognitive
        account of racial essentialism 46

evolutionary cognitive theorists endorse
    18n5
as inferred from absence of essence 23
racism
    representation as natural 94
radical constructionists 148–9, 154–6
    treat mind to world determination as
        fundamental 155–6
rape, capacity for represented as an adaptation
    97–101
rational action see action, intentional
reactive attitudes
    characterized 100, 103–4
    modified by representation as natural 73
realism 137–41
    characterized 4, 140, 164–5
    constructionism as a variety of 3, 4,
        137–61
    constructionist threats to 137
    language of "the real" 157–8
    reasons responsive 116, 131–2
    reduced attribution 95–7, 100–10
reference
    ambiguous 188
    justification of a theory of 187
    see also causal historical reference;
        descriptivism
referential grounding 186, 188–204
reflectivism 129–31
reification problem 216
representational attitudes 6–7
representational construction
    versus categories 1, 6, 47, 48
        asymmetric plausibility 9
        different threats to realism 141
    impersonal versus personal 141–2
    as a kind of category construction 49n1
Representational Control 8, 16, 54
representations
    characterized 6
    consequentialist analysis of 97–8, 107–10
    constrained by world 1–2
    as explanatory mechanisms 1–2
    as means of social regulation and control 3
    as relata of representational attitudes 6
    social-political concern with content of 3, 10,
        96–7, 101, 106–10, 112, 129,
        132–3, 210
response dependence 153–4
responsibility 113
revelation 8, 10
Revelatory Aim 8, 112
robot cats, Martian 197

Salient Possibility 71–2
"science wars" 10–11, 140
scientific practice, social accounts of 10–11

scientific revolution 17, 21, 23
segregation 87–8, 176
self-deception 121
self-explanation 125–6
self-fulfilling prophecy 80, 175
self-presentation 3
semantic individuation 50, 51–3
sex *see* gender and sex
skin color 17, 36, 56, 189, 199
  as criterial for a race concept 52
  as employed in inferences about people but not artifacts 31n25
  as indicator of race 23, 37–9
  as indicator of social group but not parents' group 36
  as input to cognitive processes in racial classification 38–9
  not a target of prejudice in ancient Mediterranean 33
  as a public broadcast 66
  similar range in Latin America and U.S. 35
social activism 172–3
social conditions of construction 57
  as common knowledge 57, 58, 59–63
social construction, in general
  characterized 16
  contingency of 5
  as explanation 1–3
  how possibly explanation of 5
  political aims of 3, 207
  sources of 4–5
social identity 115
  as an intentional response to social role 70–1
social ordering 177–8, 180–1
social role kinds 183, 195
social roles 51, 58–67, 209
  as causally insignificant 66–7
  characterized 58
  as constructionist explanations of categories 10
  entrenched 10, 11, 67, 69, 78, 93, 207, 211
  as facilitating coordination and cooperation in large groups 179–80
sortal essentialism 26n14
species concept
  classic biological concept 91–2, 200

species-typical traits 17–18, 19, 25, 25n13, 32, 42, 44, 45–6, 52, 64n16, 201
  and anti-essentialism 18n6
splitting critique of the evolutionary-cognitive account of essentialism 44
stability, of categories 11, 63, 88, 89, 92, 162–80
stereotype threat 79–80, 172
Strategic reasons 72, 73–5
strong program of the sociology of scientific knowledge 145–6
substitutionism 208
Success 137, 138, 140–1
  characterized 139
  qualified for constructionist 139–40, 157, 158–60
  constructionist threat 140–1, 143–7
  depends upon rate of theory change versus world change 166, 173
Successful Reference 8, 11, 183
switched reference 189, 191, 194–5, 202, 204
  difficulties for constructionists 195–7
symmetry in sociological explanations of theory 145–6

taking responsibility 131–2
Tandem Construction 8, 54, 55
Taylor Instability 164–7, 173
  characterized 167
  versus Hacking instability 170
thick individuation of human category 50–1
third-party classification 57
threshold of probability for asserting human category claims 108–9
tribal instincts hypothesis 77–8, 179–80
trivial reference 185, 203

ultrasociality 177–9
unconscious mental states 121–4, 129–30

Vezo identity 41–2
view from nowhere 159–60

weapon bias 79
we-belief 62
Whitehall studies 86–7
wrong kind of reason 109–10
wrong kind of referent 191, 193–4, 202